CONCEPTS OF SPACE
The History of Theories of Space in Physics

CONCEPTS OF SPACE

The History of Theories of Space in Physics

SECOND EDITION

MAX JAMMER

Foreword by Albert Einstein

HARVARD UNIVERSITY PRESS

Cambridge, Massachusetts

PREFACE TO THE SECOND EDITION

The appearance of this new enlarged and revised edition of *Concepts of Space* gives me the opportunity to express my appreciation to all those who so very kindly offered constructive criticisms on the earlier issues of the book. I am particularly grateful to Professors Mario Bunge (McGill University, Montreal), Marcus Fierz (Swiss Federal Institute of Technology, Zürich), Joseph O. Hirschfelder (University of Wisconsin, Madison), Victor F. Lenzen (University of California, Berkeley), Edward Rosen (City College, New York), as well as to the late Alexandre Koyré (Paris), whose suggestions and comments proved most valuable. I also wish to express my indebtedness to Professor Adolf Grünbaum, President of the Philosophy of Science Association, for a prolonged correspondence and for his paper in *The Philosophical Review* pertaining to some issues raised in this book. Finally, it is my pleasure to thank the National Science Foundation for a Senior Scientist Fellowship which enabled me to complete the present revision of this book while enjoying the stimulating atmosphere of the Minnesota Center for the Philosophy of Science, directed by Professor Herbert Feigl.

M. J.

Bar-Ilan University, Ramat-Gan, Israel
Columbia University, New York

PREFACE TO THE FIRST EDITION

It is my firm conviction that the study of the history of scientific thought is most essential to a full understanding of the various aspects and achievements of modern culture. Such understanding is not to be reached by dealing with the problems of priority in the history of discoveries, the details of the chronology of inventions, or even the juxtaposition of all the histories of the particular sciences. It is the history of scientific thought in its broadest perspective against the cultural background of the period which has decisive importance for the modern mind.

The concept of space, in spite of its fundamental role in physics and philosophy, has never been treated from such a historical point of view. To meet this need an attempt has been made in the following pages to present the historical development of this concept and its corresponding theories.

Although the subject has attracted my attention for a long time, it was only recently, while lecturing at Harvard University, that I found at my disposal the necessary documentary material for the writing of this book. Since I was careful to confine myself to the treatment of "space" as a concept in physics, I had to omit many theories of space that are of special interest only to the professional philosopher. However, it would have violated my principle of broad perspective had I ignored any relevant metaphysical or even theological speculations on the subject.

A presentation of the historical development of a concept does not necessarily imply adherence to a strict chronological order of discussion. A topical treatment seems to be superior for the clear crystallization of the principal ideas involved, all the more so when, as in our case, it does not seriously violate the chronological order.

Most sources from which I have drawn my information are quoted extensively, some in their original language, but the majority in English. I have also supplied abundant bibliographical references so that the interested reader can readily check my contentions and pursue the study of particular points.

I am happy to acknowledge publicly my indebtedness to Professor Albert Einstein for the great interest he has manifested in this research and for his kind provision of the foreword. I had the privilege of discussing with him at the Institute for Advanced Study many important issues of the subject. I am also indebted to Professor George Sarton, to Professor I. Bernard Cohen, and to Professor H. A. Wolfson for their valuable suggestions and helpful criticism in the early stages of the work. Others to whom my sincere thanks must be accorded are the staffs of the Widener and Houghton Libraries at Harvard University and of the Butler Library at Columbia University. Further thanks are due to the United States Department of State, for its interest in my research, to Professor Alexander Dushkin, and to all my colleagues at the Hebrew University with whom I discussed various features of the subject. In conclusion the author's gratitude is expressed to the Harvard University Press and in particular to its Science Editor, Mr. Joseph D. Elder, for the encouragement received.

M. J.

Bar-Ilan University
Ramat-Gan, Israel

CONTENTS

FOREWORD

BY ALBERT EINSTEIN

In order to appreciate fully the importance of investigations
such as the present work of Dr. Jammer one should consider the
following points. The eyes of the scientist are directed upon those
phenomena which are accessible to observation, upon their apper-
ception and conceptual formulation. In the attempt to achieve a conceptual formulation of the confusingly immense body of observational
data, the scientist makes use of a whole arsenal of concepts which he
imbibed practically with his mother's milk; and seldom if ever is he
aware of the eternally problematic character of his concepts. He uses
this conceptual material, or, speaking more exactly, these conceptual
tools of thought, as something obviously, immutably given; something
having an objective value of truth which is hardly ever, and in any
case not seriously, to be doubted. How could he do otherwise? How

would the ascent of a mountain be possible, if the use of hands, legs, and tools had to be sanctioned step by step on the basis of the science of mechanics? And yet in the interests of science it is necessary over and over again to engage in the critique of these fundamental concepts, in order that we may not unconsciously be ruled by them. This becomes evident especially in those situations involving development of ideas in which the consistent use of the traditional fundamental concepts leads us to paradoxes difficult to resolve.

Aside from the doubt arising as to the justification for the use of the concepts, that is to say, even in cases where this doubt is not in the foreground of our interest, there is a purely historical interest in the origins or the roots of the fundamental concepts. Such investigations, although purely in the field of history of thought, are nevertheless in principle not independent of attempts at a logical and psychological analysis of the basic concepts. But the limitations to the abilities and working capacity of the individual are such that we but rarely find a person who has the philological and historical training required for critical interpretation and comparison of the source material, which is spread over centuries, and who at the same time can evaluate the significance of the concepts under discussion for science as a whole. I have the impression that Dr. Jammer, through his work, has demonstrated that in his case these conditions are in great measure satisfied.

In the main he has limited himself—wisely, it seems to me—to the historical investigation of the concept of *space*. If two different authors use the words "red," "hard," or "disappointed," no one doubts that they mean approximately the same thing, because these words are connected with elementary experiences in a manner which is difficult to misinterpret. But in the case of words such as "place" or "space," whose relation with psychological experience is less direct, there exists a far-reaching uncertainty of interpretation. The historian attempts to overcome such uncertainty by comparison of the texts, and by taking into account the picture, constructed from literature, of the cultural stock of the epoch in question. The scientist of the present, however, is not primarily trained or oriented as a historian; he is not capable of forming nor willing to form his views on the origin of the fundamental concepts in this manner. He is more inclined to allow his views on the manner in which the relevant concepts might have been formed, to arise intuitively from his rudimentary knowledge of the achievements of science in the different epochs of history. He will, however, be grateful to the historian if the latter can convincingly correct such views of purely intuitive origin.

Now as to the concept of space, it seems that this was preceded by the psychologically simpler concept of place. Place is first of all a (small) portion of the earth's surface identified by a name. The thing whose "place" is being specified is a "material object" or body. Simple analysis shows "place" also to be a group of material objects. Does the word "place" have a meaning independent of this one, or can one assign such a meaning to it? If one has to give a negative answer to this question, then one is led to the view that space (or place) is a sort of order of material objects and nothing else. If the concept of space is formed and limited in this fashion, then to speak of empty space has no meaning. And because the formation of concepts has always been ruled by instinctive striving for economy, one is led quite naturally to reject the concept of empty space.

It is also possible, however, to think in a different way. Into a certain box we can place a definite number of grains of rice or of cherries, etc. It is here a question of a property of the material object "box," which property must be considered "real" in the same sense as the box itself. One can call this property the "space" of the box. There may be other boxes which in this sense have an equally large "space." This concept "space" thus achieves a meaning which is freed from any connection with a particular material object. In this way by a natural extension of "box space" one can arrive at the concept of an independent (absolute) space, unlimited in extent, in which all material objects are contained. Then a material object not situated in space is simply inconceivable; on the other hand, in the framework of this concept formation it is quite conceivable that an empty space may exist.

These two concepts of space may be contrasted as follows: (a) space as positional quality of the world of material objects; (b) space as container of all material objects. In case (a), space without a material object is inconceivable. In case (b), a material object can only be conceived as existing in space; space then appears as a reality which in a certain sense is superior to the material world. Both space concepts are free creations of the human imagination, means devised for easier comprehension of our sense experience.

These schematic considerations concern the nature of space from the geometric and from the kinematic point of view, respectively. They are in a sense reconciled with each other by Descartes' introduction of the coördinate system, although this already presupposes the logically more daring space concept (b).

The concept of space was enriched and complicated by Galileo and Newton, in that space must be introduced as the independent cause of the inertial behavior of bodies if one wishes to give the classical principle of inertia (and therewith the classical law of motion) an exact meaning. To have realized this fully and clearly is in my opinion one of Newton's greatest achievements. In contrast with Leibniz and Huygens, it was clear to Newton that the space concept (*a*) was not sufficient to serve as the foundation for the inertia principle and the law of motion. He came to this decision even though he actively shared the uneasiness which was the cause of the opposition of the other two: space is not only introduced as an independent thing apart from material objects, but also is assigned an absolute role in the whole causal structure of the theory. This role is absolute in the sense that space (as an inertial system) acts on all material objects, while these do not in turn exert any reaction on space.

The fruitfulness of Newton's system silenced these scruples for several centuries. Space of type (*b*) was generally accepted by scientists in the precise form of the inertial system, encompassing time as well. Today one would say about that memorable discussion: Newton's decision was, in the contemporary state of science, the only possible one, and particularly the only fruitful one. But the subsequent development of the problems, proceeding in a roundabout way which no one then could possibly foresee, has shown that the resistance of Leibniz and Huygens, intuitively well founded but supported by inadequate arguments, was actually justified.

It required a severe struggle to arrive at the concept of independent and absolute space, indispensable for the development of theory. It has required no less strenuous exertions subsequently to overcome this concept—a process which is probably by no means as yet completed.

Dr. Jammer's book is greatly concerned with the investigation of the status of the concept of space in ancient times and in the Middle Ages. On the basis of his studies, he is inclined toward the view that the modern concept of space of type (*b*), that is, space as container of all material objects, was not developed until after the Renaissance. It seems to me that the atomic theory of the ancients, with its atoms existing separately from each other, necessarily presupposed a space of type (*b*), while the more influential Aristotelian school tried to get along without the concept of independent (absolute) space. Dr. Jammer's views concerning theological influences on the development of the concept of space, which lie outside the range of my judgment, will certainly arouse the interest of those who are concerned with the problem of space primarily from the historical point of view.

The victory over the concept of absolute space or over that of the inertial system became possible only because the concept of the material object was gradually replaced as the fundamental concept of physics by that of the field. Under the influence of the ideas of Faraday and Maxwell the notion developed that the whole of physical reality could perhaps be represented as a field whose components depend on four space-time parameters. If the laws of this field are in general covariant, that is, are not dependent on a particular choice of coördinate system, then the introduction of an independent (absolute) space is no longer necessary. That which constitutes the spatial character of reality is then simply the four-dimensionality of the field. There is then no "empty" space, that is, there is no space without a field. Dr. Jammer's presentation also deals with the memorable roundabout way in which the difficulties of this problem were overcome, at least to a great extent. Up to the present time no one has found any method of avoiding the inertial system other than by way of the field theory.

Princeton, New Jersey
1953

CONCEPTS OF SPACE

INTRODUCTION

Space is the subject, especially in modern philosophy, of an extensive metaphysical and epistemological literature. From Descartes to Alexander and Whitehead almost every philosopher has made his theory of space one of the cornerstones of his system. The theory of relativity has led to an enormous increase in the literature on space and time. Under the influence of logical positivism the physical implications of recent theories of space have been recognized, whereas eighteenth- and nineteenth-century works were almost completely confined to purely metaphysical or psychological considerations.

Surprising as it may seem, it is a fact that no historical research on the concept of space has been published so far that deals with the history of the subject from the standpoint of physics. In the light of our modern ideas on physical space, such a treatise would be of interest not only to the historian of science and philosophy, but also to all who share in the great adventure of the intellectual progress of mankind.

It is the purpose of this monograph to show the development of the concept of space in the light of the history of physics. On the one hand the most important space conceptions in the history of scientific thought will be explained and their influence on the respective theories of mechanics and physics will be investigated; and on the other, it will be shown how experimental and observational research — together with theological speculations — affected the formulation of the corresponding metaphysical foundations of natural science as far as space is concerned. The theory of absolute space, as it finally crystallized in Newtonian mechanics, will be presented together with the criticism of it by the first modern relativists, Leibniz and Huygens. A discussion on the final elimination of the concept of absolute space from the conceptual scheme of modern physics will bring this monograph to its conclusion.

Newton's conception of absolute space is based upon a synthesis of two heterogeneous elements. One of these elements is rooted in the emancipation of space from the scholastic substance-accident scheme, a scheme which was finally abandoned by the Italian natural philosophers of the Renaissance. The other element draws on certain ideas that identify space with an attribute of God. These ideas appear to go back to Palestinian Judaism of the first century. They were adopted by Jewish mystical philosophy, and, with the spread of cabalistic teachings to Western Europe, they found an especially fruitful soil in seventeenth-century England. Under the influence of Henry More, an ardent scholar of cabalistic lore, Newton thought it necessary and expedient to make these theological ideas an integral part of his theory of space. We have, therefore, two more or less independent intellectual developments reaching back to antiquity and coming together in Newton's theory of absolute space.

Accordingly, our treatise dealing with the historical development of the concept of physical space[1] is not one continuous narrative, but is interrupted for the purpose of tracing the theological influence. So the first chapter expounds the concept of space from earliest antiquity until toward the close of Hellenistic science; the second chapter deals with the theological influences down to the time of Henry More; the third chapter resumes the subject of the first chapter; the fourth chapter deals exclusively with Newton's concept of space and Leibniz's and Huygens's criticism of it; the last chapter shows the post-Newtonian development of the concept of space and its final elimination in modern physics. In presenting the subject great care has been given to an accurate documentation of the material.

As far as pre-Newtonian and Newtonian physics are concerned we can confine our discussion to the concept of space, space and time being completely heterogeneous and noninterdependent[2] entities, although connected by the concept of motion.[3] Historically and psychologically, a discussion of space is preferable to that of time, since most probably the category of space preceded that of time as an object of consciousness. Language proves this assumption: qualifications of time, as "short," or "long," are taken from the vocabulary of spatial concepts. We say "thereafter" and not the more logical "thenafter";

[1] For an exact definition of this concept, see R. Carnap, "Der Raum. Ein Beitrag zur Wissenschaftslehre," *Kantstudien,* Ergänzungsheft No. 56 (1922).

[2] In the Galilean transformation of classical mechanics, $t' = t$, that is, the transformed time variable is independent of the space variable.

[3] As pointed out by C. A. Brandis in his *Griechisch-römische Philosophie* (Berlin, 1835), vol. 1, pp. 413, 415, Zeno of Elea seems to have been the first who emphasized this connection between space and time. Cf. Locke's *Essay concerning human understanding* (London, 1785), vol. 1, pp. 149, 156: ". . . to measure motion space is as necessary to be considered as time . . . They are made use of, to denote the position of finite real Beings in respect one to another in those uniform oceans of Duration and Space."

"always" means "at all times"; we even speak of a "space" or an "interval" of time: "before" means etymologically "in front of." In this respect the Semitic languages are especially instructive, a fact pointed out by Ignaz Goldziher.[4] The Hebrew word for "before" is "lifney," which originally means "to the face of," "to the front of"; many other words, for instance "Kedem," "aharey," show clearly the trend from spatial to temporal qualifications. As a matter of fact, this trend can be recognized already in the ancient Sumerian expression *danna,* which was originally a measure of length and later signified a certain fraction of the day (unit of time).[5] Modern psychology undoubtedly allows more concreteness to the concept of space than to the concept of time. If we remember that it was only late in the Middle Ages that the role of time as the fundamental variable parameter in physical processes was clearly understood, we can justify our concentration on the concept of space, at least as far as early theories of space are concerned.

But we are fully aware of the fact that since Leibniz's profound analysis of the concepts of space and time the notion of time has often been held to precede the notion of space in the construction of a philosophical system. The direction of the flow of time was thought to be determined by the causal interconnection of phenomena. Space, then, was only the order of coexisting data. "Spatium est ordo coexistendi," said Leibniz in his *Initia rerum metaphysica,* a surprisingly modern analysis of our concepts of space and time.

Similarly, some modern philosophers of science, in their attempt to establish deductively the structure of space-time, base their investigation on the notion of temporal order and try to derive from it the topological properties of space. Thus, for example, Carnap's profound study[6] of 1925 on the dependence of the properties of

[4] Ignaz Goldziher, *Mythology among the Hebrews* (London, 1877).

[5] O. Neugebauer, "Untersuchungen zur Geschichte der antiken Astronomie, III," *Quellen und Studien zur Geschichte der Mathematik* (Springer, Berlin, 1938), part B, vol. 4, p. 193.

[6] R. Carnap, "Über die Abhängigkeit der Eigenschaften des Raumes von denen der Zeit," *Kantstudien 30,* 331–345 (1925).

space on those of time was based exclusively on the following two relations: (1) spatio-temporal coincidence and (2) temporal precedence among world-points (in the sense of Minkowski). Reichenbach in his systematic study of space and time similarly claimed that space measurements are reducible to time measurements. In fact, he stated explicitly: "Time is . . . logically prior to space." [7] Another attempt to derive the Minkowski metric of space from purely temporal relations was made by Markoff[8] on the assumption of a discrete structure of time and by means of a special *ad hoc* axiom, the "Viereckaxiom," whose function it is to distend time into space. The axiomatic space-time theory of Robb[9] and the well-known cosmological system of Milne[10] claim that the metrical structure of space-time can be established purely on the basis of the use of light signals and the relation of temporal succession. One of the most eloquent proponents of this view at present probably is Synge, who unhesitatingly proclaims: "Euclid put us on the wrong track, so that we put space first and time second — a very poor second indeed." [11] Finally, also in Basri's[12] recently published theory of space and time — in spite of the order in which these concepts appear in the title of his book — time precedes space in the order of constructing the foundations of theoretical physics.

All these attempts to derive spatiality or extension from pure temporality, conceived as a one-dimensional order of succession, seem, however, to be open to two serious objections: (1) The use of light signals and temporal succession without an assumption of the existence of rigid rods or material clocks (and hence spatially extended objects) is insufficient for the measurement of spatial in-

[7] H. Reichenbach, *The philosophy of space and time* (Dover Publications, New York, 1958), p. 169.
[8] A. Markoff, "Über die Ableitbarkeit der Weltmetrik aus der 'Früher Als' Beziehung," *Physikalische Zeitschrift der Sowjetunion 1*, 397–406 (1932).
[9] A. A. Robb, *A theory of time and space* (Cambridge University Press, Cambridge, Eng., 1913, 1914, 1936).
[10] E. Milne, *Kinematic relativity* (Oxford University Press, London, 1948).
[11] J. L. Synge, "A plea for chronometry," *The new scientist* (February 19, 1959), pp. 410–412.
[12] S. Basri, *A deductive theory of space and time* (North-Holland Publishing Company, Amsterdam, 1966).

tervals, an argument pointed out already by Whyte[13] in 1954. (2)
The very admission of a multiplicity of world-lines presupposes,
even if only in a rudimentary form, some kind of spatiality. Only
if time may be regarded, not as a one-dimensional continuum of
instants as conceived in the classical way, but rather as being endowed with a certain transversal extent, as intimated by Čapek,[14]
who followed in this context Bergson's philosophy of extensive becoming and Whitehead's idea of the creative advance of nature —
only then does it seem to be possible to derive spatiality from
temporality. But these and similar metaphysical conceptions have
not yet been absorbed by science: Geometry, in the sense of a
science of space, has not yet been logically subordinated to chronometry, the science of time and its measurement. Finally, as far
as classical conceptions of space are concerned, we may safely
regard the concept of space as an elementary and primary notion.

[13] "A physicist using only light signals cannot discriminate inertial systems
from these subjected to arbitrary similarity transformations. The system of
'resting' mass-points which can so be identified may be arbitrarily expanding
and/or contracting relative to a rod, and these superfluous transformations can
only be eliminated by using a rod or a clock." L. L. Whyte, "Light signal
kinematics," *The British journal for the philosophy of science 4*, 160–161
(1954). Ultimately, the reason for this objection is grounded in the fact that
the Maxwell equations, as H. Bateman and E. Cunningham had shown in
1910, are invariant under the conformal group of transformations in four-
dimensional Minkowski space, a group which also includes, in addition to
translations, rotations, and reflections, inversions with respect to the hyper-
spheres of this space and hence transformations which change inertial frames
of reference into frames of reference that are not inertial.
[14] M. Čapek, *The philosophical impact of contemporary physics* (D. Van
Nostrand Company, Princeton, 1961).

CHAPTER 1

THE CONCEPT OF SPACE

IN ANTIQUITY

Modern physics on the whole — if we neglect certain relativistic theories — qualifies space as continuous, isotropic, homogeneous, finite, or infinite, in so far as it is not a pure system of relations. Not all of these qualities, however, are accessible to sense perception. They are the result of a long and continuous process of abstraction which had its beginning in the mind of primitive man. Philological, archaeological, and anthropological research shows clearly that primitive thought was not capable of abstracting the concept of space from the experience of space. To the primitive mind, "space" was merely an accidental set of concrete orientations, a more or less ordered

multitude of local directions, each associated with certain emotional reminiscences. This primitive "space," as experienced and
subconsciously formed by the individual, may have been co-
ordinated with a "space" common to the group, the family or
the tribe. Certain astronomical or meteorological events, such as
sunrise and sunset, storms and floods, no doubt endowed certain directions with values of common importance. "Mesopota-
mian astrology evolved a very extensive system of correlations
between heavenly bodies and events in the sky and earthly
localities. Thus mythopoeic thought may succeed no less than
modern thought in establishing a coördinated spatial system; but
the system is determined, not by objective measurements, but
by an emotional recognition of values." [1] It can be shown that
even with the introduction of conventional standards of measurement in early urban society, lengths, areas, and volumes were
not conceived *in abstracto* as purely spatial extensions. To be
sure, measurement leads eventually to generalization and ultimately to abstract thinking. Ignoring the color, design, and tex-
ture of the object to be measured, human thought begins by
"abstraction" to concentrate on the idea of pure extension and
unqualified space. However, it must not be supposed that this
was a simple and short process. Archaeology shows that the early
abstractions were limited by practical interests. The ancient Sumerian unit of area — incidentally also the unit of weight — was
the *še* or "grain." This designation indicates clearly that areal extension was in those times conceived from the aspect of the
quantity of seed necessary for the sowing of the area in question,
which means, in the final analysis, from the anthropocentric
aspect of the labor involved.

Hesiod's "chaos," [2] which may be taken as the earliest poetical
expression of the idea of a universal space, is mixed with emotion;

[1] H. Frankfort, H. A. Frankfort, J. A. Wilson, and T. Jacobson, *The intellectual adventure of ancient man* (University of Chicago Press, Chicago, 1946), p. 20.
[2] Hesiod, *Theogony,* 116. Cf. Deichmann's objection to Zeller's interpretation in Carl Deichmann, *Das Problem des Raumes in der griechischen Philosophie bis Aristoteles* (Halle a. S., 1893).

the very word "chaos," derived from the Greek root *cha-* (*cha-skein, chainein*), implies as "yawning," "gaping," an idea of terror and fright. To what extent such poetical-mystical concepts have been conditioned by early folklore and myth (such as the Aditi lore of the Arians) is a matter that falls outside the scope of this monograph.

Space as a subject of philosophical inquiry appears very early in Greek philosophy. According to Aristotle,[3] numbers were accredited with a kind of spatiality by the Pythagoreans: "The Pythagoreans, too, asserted the existence of the void and declared that it enters into the heavens out of the limitless breath — regarding the heavens as breathing the very vacancy — which vacancy 'distinguishes' natural objects, as constituting a kind of separation and division between things next to each other, its prime seat being in numbers, since it is this void that delimits their nature." Spatial vacancies were necessary to guarantee the discreteness of individual numbers in the Pythagorean geometrization of number. Space here has not yet any physical implications apart from serving as the limiting agent between different bodies. In early Pythagorean philosophy this kind of "space" is still called *pneuma apeiron* and only occasionally *kenon* (void). The concept of space is still confounded with that of matter. As J. Burnet says: "The Pythagoreans, or some of them, certainly identified 'air' with the void. This is the beginning, but no more than the beginning, of the conception of abstract space or extension."[4] Only later on is this confusion cleared up by Xutus and Philolaus.[5] In Simplicius[6] we find that Archytas, the Pythagorean, already had a clear understanding of this abstract notion, since, as related by Eudemus, he asked whether it would be possible at the end of the world to stretch out one's hand or not. Unfortunately, Archytas' work on the nature of space is lost except

[3] Aristotle, *Metaphysics*, 1080 b 33.
[4] J. Burnet, *Early Greek philosophy* (London, 1914), p. 51.
[5] P. Tannery, *Revue philosophique* 20 (1885), 389.
[6] Simplicius, *Physics*, 108 a.
[7] *In Aristotelis categorias commentarium* (ed. Carolus Kalbfleisch; Berolini, 1907), p. 13.

for a few fragments to be found in Simplicius' *Commentaries*,[7] according to which Archytas composed a book on our subject. Archytas distinguishes between place (*topos*), or space, and matter. Space differs from matter and is independent of it. Every body occupies some place, and cannot exist unless its place exists. "Since what is moved is moved into a certain place and doing and suffering are motions, it is plain that place, in which what is done and suffered exists, is the first of things. Since everything which is moved is moved into a certain place, it is plain that the place where the thing moving or being moved shall be, must exist first. Perhaps it is the first of all beings, since everything that exists is in a place and cannot exist without a place. If place has existence in itself and is independent of bodies, then, as Archytas seems to mean, place determines the volume of bodies." [8] A characteristic property of space is that all things are in it, but it is never in something else; its surroundings are the infinite void itself. Apart from this metaphysical property, space has the physical property of setting frontiers or limits to bodies in it and of preventing these bodies from becoming indefinitely large or small. It is also owing to this constraining power of space that the universe as a whole occupies a finite space. To Archytas, space is therefore not some pure extension, lacking all qualities or force, but is rather a kind of primordial atmosphere, endowed with pressure and tension and bounded by the infinite void.

The function of the void, or of space, in the atomism of Democritus is too well known to need any elaboration here. But it is of interest to note that according to Democritus infinity of space is not only inherent in the concept itself,[9] but may be deduced from the infinite number of atoms in existence, since these, although indivisible, have a certain magnitude and extension, even if they are not perceptible to our senses. Democritus himself seems not to have attributed weight to the atoms but to have assumed that as a result of constant collisions among them-

[8] *Ibid.*, p. 357.
[9] Aristotle, *De caelo*, III, 2, 300 b.

selves they were in motion in infinite space. It was only later, when an explanation of the cause of their motion was sought, that his disciples introduced weight as the cause of the "up and down" movements (Epicurus). If Aristotle says that Democritus' atoms differed in weight according to their size, one has to assume — in modern words — that it was not gravitational force but "force of impact" that was implied. This point is of some importance for our point of view, since it shows that in the first atomistic conception of physical reality space was conceived as an empty extension without any influence on the motion of matter.

However, there still remains one question to be asked: Was space conceived by the atomists of antiquity as an unbounded extension, permeated by all bodies and permeating all bodies, or was it only the sum total of all the *diastemata*, the intervals that separate atom from atom and body from body, assuring their discreteness and possibility of motion? The stress laid time and again by the atomists on the existence of the void was directed against the school of Parmenides and Melissus, according to whom the universe was a compact plenum, one continuous unchanging whole. "Nor is there anything empty," says Melissus, "for the empty is nothing and that which is nothing cannot be." Against such argument Leucippus and Democritus maintained the existence of the void as a logical conclusion of the assumption of the atomistic structure of reality. But here the void or the empty means clearly unoccupied space. The universe is the full and the empty. Space, in this sense, is complementary to matter and is bounded by matter; matter and space are mutually exclusive. This interpretation gains additional weight if we note that the term "the empty" (*kenon*) was used often as synonymous with the word "space"; the term "the empty" obviously implies only the unoccupied space. Additional evidence is furnished by Leucippus' explicit use of the adjective "porous" (*manon*) for the description of the structure of space, which indicates that he had in mind the intervals between particles of matter and not unbounded space. Although Epicurus' recurrent description of the

universe as "body and void" seems also to confirm this interpretation, we find in Lucretius, who bases himself on Epicurus,
a different view. In general, Lucretius' complete and coherent
scheme of atomistic natural philosophy is the best representation
of Epicurean views. As far as the problem of space is concerned,
Lucretius emphasizes in the first book of *De rerum natura* the
maxim: "All nature then, as it exists, by itself, is founded on two
things: there are bodies and there is void in which these bodies
are placed and through which they move about." [10]

Here we find, in contrast to the early Greek atomism, a clear
and explicit expression of the idea that bodies are placed in the
void, in space. With Lucretius, therefore, space becomes an
infinite receptacle for bodies. Lucretius' proof for the unbounded-
ness of space, resembling Archytas' argument mentioned earlier,[11]
runs as follows: "Now since we must admit that there is nothing
outside the sum, it has no outside, and therefore is without end
and limit. And it matters not in which of its regions you take
your stand; so invariably, whatever position any one has taken
up, he leaves the universe just as infinite as before in all directions. Again, if for the moment all existing space be held to be
bounded, supposing a man runs forward to its outside borders
and stands on the utmost verge and then throws a winged javelin,
do you choose that when hurled with vigorous force it shall advance to the point to which it has been sent and fly to a distance,
or do you decide that something can get in its way and stop it?
for you must admit and adopt one of the two suppositions; either
of which shuts you out from all escape and compels you to grant
that the universe stretches without end." [12]

[10] T. Lucreti Cari, *De rerum natura* (trans. by Munro; Cambridge, 1886),
vol. 3, p. 23. The original Latin text is:

. . . nam corpora sunt et inane,
haec in quo sita sunt et qua diversa moventur.
— Liber I, 420.

[11] See p. 8.
[12] Reference 10.

This argument, and in particular the idea of a man placed at the supposed boundary of space stretching out his hand or throwing a spear, is a recurrent idea in the history of natural philosophy. In fact, an illustration of this kind is to be expected. We find it in Richard of Middleton's writings[13] in the fourteenth century (perhaps with reference to Simplicius' *Physics* 108 a), still before the rediscovery of the *De rerum natura* in 1418 by Poggio. We also find it as late as in Locke's *Essay concerning human understanding* (1690), where the question is asked "whether if God placed a man at the extremity of corporeal beings, he could not stretch his hand beyond his body."[14]

Lucretius adduces a further argument for the infinitude of space which reveals an important physical aspect of the atomistic theory: If space were not infinite, he claims, all matter would have sunk in the course of past eternity in a mass to the bottom[15] of space and nothing would exist any more. This remark shows clearly that Lucretius, in the wake of Epicurus, conceived space as endowed with an objectively distinguished direction, the vertical. It is in this direction in which the atoms are racing through space in parallel lines. According to Epicurus and Lucretius, space, though homogeneous, is not isotropic.

Although the idea of a continuous homogeneous and isotropic space, as we see, seems to have been too abstract even for the theoretically minded atomists, it has been justly pointed out[16] that their conception of the noncorporeal existence of a void introduced a new conception of reality. Indeed, it is a strange coincidence that the very founders of the great materialistic school in antiquity had to be "the first to say distinctly that a thing might be real without being a body."

The first clear idea of space and matter as belonging to dif-

[13] See chapter 3, reference 46.
[14] John Locke, *An essay concerning human understanding*, book II, 13, 21; see, for example, the edition by A. S. Pringle-Pattison (Clarendon Press, Oxford, 1950), p. 102.
[15] *Ad imum.* Liber I, 987.
[16] J. Burnet, *Early Greek philosophy* (London, ed. 3, 1920), p. 389.

ferent categories is to be found in Gorgias.[17] Gorgias first proves that space cannot be infinite. For if the existent were infinite, it would be nowhere. For were it anywhere, that wherein it would be, would be different from it, and therefore the existent, encompassed by something, ceases to be infinite; for the encompassing is larger than the encompassed, and nothing can be larger than the infinite; therefore the infinite is not anywhere. Nor on the other hand, can it be encompassed by itself. For in that case, that wherein it is found would be identical with that which is found therein, and the existent would become two things at a time, space and matter; but this is impossible. The impossibility of the existence of the infinite excludes the possibility of infinite space.

Plato, who, according to Aristotle, was not satisfied, as his predecessors were, with the mere statement of the existence of space, but "attempted to tell us what it is," [18] develops his theory of space mainly in *Timaeus*. The upshot of the rather obscure exposition of this dialogue, as interpreted by Aristotle,[19] and in modern times by E. Zeller,[20] is that matter — at least in one sense of the word — has to be identified with empty space. Although "Platonic matter" was sometimes held to be a kind of body lacking all quality (Stoics, Plutarch, Hegel) or to be the mere possibility of corporeality (Chalcidius, Neoplatonists), critical analysis seems to show that Plato intended to identify the world of physical bodies with the world of geometric forms. A physical body is merely a part of space limited by geometric surfaces containing nothing but empty space.[21] With Plato physics becomes geometry, just as with the Pythagoreans it became arithmetic. Stereometric similarity becomes the ordering principle in the formation of macroscopic bodies. "Now the

[17] *Sexti Empirici opera*, "Adversus dogmaticos" (ed. H. Mutschmann; Leipzig, 1912–14), vol. 2, p. 17.
[18] Aristotle, *Physics*, 209 b.
[19] *Ibid.*, 203 a, 209 b.
[20] E. Zeller, *Die Philosophie der Griechen* (Leipzig, 1869–1879), vol. 2.
[21] Plato, *Timaeus*, 55 ff.

Nurse of Becoming, being made watery and fiery and receiving the characters of earth and air, and qualified by all the other affections that go with these, had every sort of diverse appearance to the sight; but because it was filled with powers that were neither alike nor evenly balanced, there was no equipoise in any region of it; but it was everywhere swayed unevenly and shaken by these things, and by its motion shook them in turn. And they, being thus moved, were perpetually being separated and carried in different directions; just as when things are shaken and winnowed by means of winnowing-baskets and other instruments for cleaning corn, the dense and heavy things go one way, while the rare and light are carried to another place and settle there. In the same way at that time the four kinds were shaken by the Recipient, which itself was in motion like an instrument for shaking, and it separated the most unlike kinds farthest apart from one another, and thrust most alike closest together; whereby the different kinds came to have different regions, even before the ordered whole consisting of them came to be." [22] Physical coherence, or, if one likes, chemical affinity, is the outcome of stereometric formation in empty space, which itself is the undifferentiated material substrate, the raw material for the Demiurgus. The shaking and the winnowing process characterizes space with a certain stratification and anisotropy which is manifested physically in the difference between the layers of the elements. Geometric structure is the final cause of what has been called "selective gravitation," where like attracts like.

In accordance with certain ideas expressed by the Pythagorean Philolaus,[23] Plato conceived the elements as endowed with definite spatial structures:[24] to water he assigned the spatial structure of an icosahedron, to air of an octahedron, to fire of a pyramid, and to earth of a cube. Earth, in Plato's view,

[22] *Ibid.*, 52 d; F. M. Cornford, *Plato's cosmology* (Harcourt, Brace, New York, 1937), p. 198.
[23] Zeller, *Philosophie der Griechen*, vol. 1, p. 376.
[24] Plato, *Timaeus*, 56.

owing to its cubical form, is the most immovable of the four, having the most stable bases. It is only natural, therefore, that this element is found at the center of the universe; like a nucleus it is embedded in layers of the other elements of space according to their increasing movability. The varieties of the four elements and their giavitational behavior are due to differences in their form and size, or, in the final analysis, are due to differences in form and size of the elementary triangles of which their plane surfaces are formed. As much as matter is reduced to space, physics is reduced to geometry.

This identification of space and matter, or, in the words of later pseudo-Platonic teachings, of tridimensionality and matter, had a great influence on physical thought during the Middle Ages. For although Aristotle's *Organon* was the standard text in logic, Plato's *Timaeus* was succeeded by Aristotle's *Physics* only in the middle of the twelfth century. It is perhaps not wrong to assume that the obscure and vague language of the *Timaeus* contributed to preventing the concept of space from becoming a subject of strict mathematical research. Greek mathematics disregards the geometry of space. Plato himself, for whom solid bodies and their geometry were of fundamental importance in the formulation of his philosophy, lamented the neglecting of this branch of mathematics. In the *Republic*[25] he apologizes for failing to discuss solid geometry when listing the essential subjects for instruction. So we read:

Glaucon: It is true that they possess an extraordinary attractiveness and charm. But explain more clearly what you were just speaking of. The investigation of plane surfaces, I presume, you took to be geometry?
Socrates: Yes.
Glaucon: And then at first you took astronomy next and then you drew back.
Socrates: Yes, for in my haste to be done I was making less speed. For while the next thing in order is the study of the third dimension or solids, I passed it over because of our absurd neglect to investigate

[25] Plato, *Republic,* 528.

it, and mentioned next after geometry, astronomy, which deals with the movements of solids.[26]

Aristotle's theory of space is expounded chiefly in his *Categories* and, what is of greater relevance for our purpose, in his *Physics*. In the *Categories* Aristotle begins his short discussion with the remark that quantity is either discrete or continuous. "Space," belonging to the category of quantity, is a continuous quantity. "For the parts of a solid occupy a certain space, and these have a common boundary; it follows that the parts of space also, which are occupied by the parts of the solid, have the same common boundary as the parts of the solid. Thus, not only time, but space also, is a continuous quantity, for its parts have a common boundary." [27] "Space" here is conceived as the sum total of all places occupied by bodies, and "place" (*topos*), conversely, is conceived as that part of space whose limits coincide with the limits of the occupying body.[28]

In the *Physics* Aristotle uses exclusively the term "place" (*topos*), so that strictly speaking the *Physics* does not advance a theory of space at all, but only a theory of place or a theory of positions in space. However, since the Platonic and Democritian conceptions of space are unacceptable to the Aristotelian system of thought, and since the notion of empty space is incompatible with his physics, Aristotle develops only a theory of positions in space, with the exclusion of the rejected conception of general space.

For our purpose, Aristotle's theory of places is of greatest pertinence not only because of its important implications for physics, but also because it was the most decisive stage for the further development of space theories. In our treatment we shall adhere as much as possible to Aristotle's original terminology and use the term "place."

[26] Plato, *Republic*, book VII, trans. by P. Shorey (Loeb Classical Library; Harvard University Press, Cambridge, 1946), vol. 2, p. 179.
[27] Aristotle, *Categories*, 5 a, 8–14. See Richard McKeon, *The basic works of Aristotle* (Random House, New York, 1941), p. 15.
[28] For this interpretation, see Pierre Duhem, *Le système du monde* (Paris, 1913–1917), vol. 1, p. 197.

In Book IV of the *Physics* Aristotle develops on an axiomatic basis a deductive theory of the characteristics of place. Place is an accidens, having real existence, but not independent existence in the sense of a substantial being. Aristotle's four primary assumptions regarding our concept are as follows: "(1) That the place of a thing is no part or factor of the thing itself, but is that which embraces it; (2) that the immediate or "proper" place of a thing is neither smaller nor greater than the thing itself; (3) that the place where the thing is can be quitted by it, and is therefore separable from it; and lastly (4): that any and every place implies and involves the correlatives of 'above' and 'below,' and that all the elemental substances have a natural tendency to move towards their own special places, or to rest in them when there — such movement being 'upward' or 'downward,' and such rest 'above' or 'below.'" [29] It is this last assumption that makes space a carrier of qualitative differences and furnishes thereby the metaphysical foundation of the mechanics of "natural" motion. Starting from these assumptions, Aristotle proceeds by a lucid process of logical elimination[30] to his famous definition of "place" as the adjacent boundary of the containing body. By this definition the concept became immune to all the criticisms that were designed to show the logical inconsistency of former definitions, as, for instance, Zeno's famous epicheirema (Everything is in place; this means that it is in something; but if place is something, then place itself is in something, etc.). In fact, this "nest of superimposed places" is mentioned as an argument against the existence of a kind of dimensional entity — distinct from the body that has shifted away when the encircled content is taken out and changed again and again, while the encircling continent remains unchanged.

Further, this "replacement" of the content of a vessel by another content reveals that place is something different from

---

[29] Aristotle, *Physics*, 211 a, trans. by P. H. Wicksteed and F. M. Cornford (Loeb Classical Library; Harvard University Press, Cambridge, 1929), vol. 1, p. 303.
[30] *Ibid.*, 211 b.

its changing contents and so proves the reality of space. Of great importance from our point of view is a passage in Aristotle's *Physics* in which space is likened (using a modern expression) to a field of force: "Moreover the trends of the physical elements (fire, earth, and the rest) show not only that locality or place is a reality but also that it exerts an active influence; for fire and earth are borne, the one upwards and the other downwards, if unimpeded, each towards its own 'place,' and these terms — 'up' and 'down' I mean, and the rest of the six dimensional directions — indicate subdivisions or distinct classes of positions or places in general." [31]

The dynamical field structure, inherent in space, is conditioned by the geometric structure of space as a whole. Space, as defined by Aristotle, namely, as the inner boundary of the containing receptacle, is, so to speak, a reference system which generally is of very limited scope. The place of the sailor is in the boat, the boat itself is in the river, and the river is in the river bed. This last receptacle is at rest relative to the earth and therefore also to the universe as a whole, according to contemporary cosmology. For astronomy, with its moving spheres, the reference system has to be generalized still further, leading to the finite space of the universe limited by the interior boundary of the outermost sphere, which itself is not contained in any further receptacle. This universal space, of spherical symmetry, has as its center the center of the earth, to which heavy bodies move under the dynamic influence intrinsic to space. It is natural for us, who have read Mach and Einstein, to raise the question whether the geometric aspect of this dynamical "field structure" depends on the distribution of matter in space or is completely independent of mass. Aristotle anticipated this question and tried to show that the dynamics of natural motion depends on spatial conditions only.

It might be asked, since the center of both (i.e., the earth and the universe) is the same point, in which capacity the natural motion of

[31] *Ibid.*, 208 b; Loeb edition, p. 279.

heavy bodies, or parts of the earth, is directed towards it; whether as center of the universe or of the earth. But it must be towards the center of the universe that they move, seeing that light bodies like fire, whose motion is contrary to that of the heavy, move to the extremity of the region which surrounds the center. It so happens that the earth and the universe have the same center, for the heavy bodies do move also towards the center of the earth, yet only incidentally, because it has its center at the center of the universe." [32]

This description is suggestive of the electrostatic field that exists between a small charged sphere enclosed by another sphere at a different potential. As is well known, the field itself may be nonspherically symmetric, as in the case of an excentric position of the inner sphere, which corresponds to the earth when shifted from the center of the universe, although the lines of force leave the surface of the enclosed body in a normal direction. To Aristotle, such a distortion seemed to be absurd; his world is a world of order and symmetry.

The directional tendencies of the elemental particles are possible only because of the difference in the conditions of the place in which they move from the conditions of the place to which they move. It is clear, therefore, that it is not a kind of buoyancy (corresponding to Archimedes' principle) that causes the motion of heavy or light bodies. For in this case the dynamical field structure would be mass-dependent. However, although these tendencies are independent of the distribution of mass, they are dependent on the very existence of matter. A void, conceived by Aristotle as the privation of all conceivable properties, cannot by its very definition be something differentiated directionally. It is well known how Aristotle exploited this argument in his repudiation of the void.

In conformity with the rejection of a vacuum, Aristotle insists repeatedly that the containing body has to be everywhere in contact with the contained. Polemizing against the Pythagorean doctrine of spatial vacancies, Aristotle offers a psychological explanation of the origin of such "gap" theories. "Because the

---

[32] Aristotle, *De caelo*, II, 14, 296 b; Loeb edition, p. 243.

encircled content may be taken out and changed again and again, while the encircling continent remains unchanged — as when water passes out of a vessel — the imagination pictures a kind of dimensional entity left there, distinct from the body that has shifted away." [33] But, he holds, to suppose that this "interval" is the place or space of the contained would inescapably lead to serious inconsistencies. He argues that on the basis of such a "gap" theory "place" would have to change its "place" and an ascending series of orders of spaces would be involved. Thus, when carrying a vessel of water from one place to another, one has to carry about also the "interval" and a transport of space in space is implied. His second objection is based on the assertion that transporting a vessel full of water means changing the place of the whole but not the places of its parts. According to Simplicius, Aristotle's line of thought seems to have used the following *reductio ad absurdum:* On the basis of an "interval" theory every part of water has to have its own place, since a transport of a vessel of water is accompanied by a rotation or wave disturbance of the liquid, which is possible only if the parts can shift from one interval to another. However, matter is indefinitely divisible and the number of such intervals must consequently be unlimited even for the smallest quantity of water. It follows that the volume, the sum total of all these intervals, being a sum of an infinite series, is infinitely great.

While expounding the inadequacy of these "interval" theories, Aristotle, on his part, ignores the fact that his very insistence on the all-over contact of the two distinct surfaces of the container and the contained must necessarily lead to a serious inconsistency between his own space theory on the one hand and his cosmology and theology on the other. For if the interior concave surface of the sphere of one planet is everywhere in contact with the convex surface of the sphere of another, then obviously the "fifth body," the substance of which the heavens are made, is not continuous, a conclusion that is contrary to the results of his cosmological

[33] Aristotle, *Physics,* IV, 211 b 15; Loeb edition, p. 309.

doctrines as presented in *De caelo*.[34] Simplicius, who noticed this inconsistency, tried to avoid it by maintaining that all celestial spheres extend to one common center which coincides with the center of the earth. But obviously, Simplicius' solution of the problem is not only a theory *ad hoc*, but is also incompatible with the principles of Aristotelian physics which explicitly rejects the interpenetrability of different bodies.[35]

It should be noted that Aristotle's remarks in the *Categories* indicate a different way of attacking the problem of space. Here space seems to be some kind of continuous extension; it is given no strict definition, and, what is more important for our point of view, it has no physical implications for Aristotle's natural philosophy or that of his successors.

It is evident that space as an accident of matter is, according to Aristotle, finite, matter being itself finite.

Space, here, means the sum total of all places. The idea of a finite physical space, thus understood, is not as absurd today as it must have appeared fifty years ago, when physics acknowledged solely the conception of an infinite Euclidean space and when a finite material universe could but be conceived as an island, so to speak, in the infinite ocean of space. It is perhaps not wholly unjustified to suggest a comparison between the notion of physical space in Aristotle's cosmology and the notion of Einstein's "spherical space" as expounded in early relativistic cosmology. In both theories a question of what is "outside" finite space is nonsensical. Furthermore, the idea of "geodesic lines," determined by the geometry of space, and their importance for the description of the paths of material particles or light rays, suggest a certain analogy to the notion of "natural places" and the paths leading to them. The difference is, of course, that in Einstein's theory the geometry of space itself is a function of the mass-energy distribution in accord with the famous field equations, and is not Euclidean but Riemannian.

[34] Aristotle, *De caelo*, I, 3, 270 a *et seq.*, Loeb edition, p. 21.
[35] Aristotle, *Physics*, IV, 209 a 7; Loeb edition, p. 283.

Although until the fourteenth century Aristotle's and Plato's conceptions were the prototypes, with only minor changes, of all theories of space, yet these conceptions were the object of constant attack, mostly on metaphysical grounds. Aristotle's pupil Theophrastus criticizes the master's theory[36] and speaks of the possibility of a motion of space, of the incomprehensibility of the universe as not being in space, and comes to the conclusion[37] that space is no entity in itself but only an ordering relation that holds between bodies and determines their relative positions. Like a biologist who dissects an animal and considers one organ in relation to another, so Theophrastus views space as a system of interconnected relations.

Concerned as we are with the problem of space in its implications for physics, we may disregard the few original contributions of the Epicureans, Skeptics, and other schools. We should, however, mention in this connection the important deviation of the Stoics from the traditional Aristotelian conception of the cosmos. Continuity, which for Aristotle was a purely geometric property of coherent matter, became with the Stoa a physical principle, an agent responsible for the propagation of physical processes through space. It is by this inner connection, manifested as a tension (*tonos*) in its active state, that distant parts of the universe are able to influence each other, thereby turning the cosmos into one field of action. The void, being incorporeal and therefore lacking all continuity, necessarily precludes all sense perception and so cannot exist inside the world.[38]

This elaboration of the Aristotelian idea of tendencies permeating the continuous plenum is an important generalization in two respects; in the variety of phenomena envisaged, and in their extension beyond the sublunar world (for instance, Posidonius' discovery of the moon's "influence" on the tides, which was regarded as an ostensible proof of the reality of this transmitting

[36] Simplicius, *Physics*, 141.
[37] *Ibid.*, 141, 149.
[38] Cleomedes, "De motu circulari corporum caelestium libri duo," in J. ab Arnim, ed., *Stoicorum veterum fragmenta*, II, 546 (Leipzig, 1905), p. 172.

agent connecting even celestial with mundane phenomena [Chrysippus]). The range of activity of the propagating tensions is the whole material universe (*holon*) as distinct from the "All" (*pan*).

In order to explain this important distinction we have to refer to the changed definition of "space." In general, the Stoics accepted not Aristotle's definition of space, as the containing surface of the encircling body, but his discarded alternative, that is, the dimensional extension lying between the points of the containing surface. This change enabled the Stoics to maintain the existence of a void outside the material universe, whereas the material universe was conceived as an island of continuous matter surrounded by an infinite void. Needless to say, this infinite void lacked all qualities and differentiations, and, being thus completely indeterminate, it could not act in any way on the matter surrounded by it.[39] Hence the position of bodies was not determined by any properties of the void, but by their own nature. With no reason to move as a whole, the material world rests immovable in the infinite void. To the Stoics it made no sense to speak of the center of the "All"; on the other hand, the center of the material universe was a clear concept, cosmologically and physically well founded. Criticizing this doctrine, the Peripatetics raised the following question: If the material world is really surrounded by an infinite void, why does it not become dissipated and lost in the course of time?[40] The answer is now clear: the different parts of the material world are connected, not, as Aristotle thought, by an exterior continent, an upper sphere which forces the parts to stay together, like samples in a box, but by an internal cohesion (*hexis*), which is only another aspect of the tension mentioned before. It is this binding force that holds the world together, and the void without force of its own can do nothing to loosen it. In the void there is no "up" or "down" or

[39] *Ibid.*, II, 173, 176 (pp. 49, 51).
[40] *Ibid.*, II, 540.

any other direction or dimension.[41] In other words, it is isotropic, bare of any qualities whatever. As for space in the material universe, the Stoics adhere to the traditional Aristotelian doctrine. It was owing chiefly to these Stoic controversies that the problem of space could no longer be considered as one simple question, but had to appear under the form of two different considerations: space and void.

As we have tried to show in this chapter, space was conceived by classical Greek philosophy and science at first as something inhomogeneous because of its local geometric variance (as with Plato), and later as something anisotropic owing to directional differentiation in the substratum (Aristotle). It is perhaps not too conjectural to assume that these doctrines concerning the nature of space account for the failure of mathematics, especially geometry, to deal with space as a subject of scientific inquiry. Perhaps this is the reason why Greek geometry was so much confined to the plane. It may be objected that "space" according to Aristotle is "the adjacent boundary of the containing body" and so by its very definition is only of a two-dimensional character. But this objection ignores a clear passage in the *Physics*[42] and another passage in *De caelo*.[43] As Euclid's *Elements* show, the science of solid geometry was developed only to a small extent and mostly confined to the mensuration of solid bodies, which is at least one reason why even the *termini technici* of solid geometry, compared with those of plane geometry, were so little standardized. The idea of coördinates in the plane seems to go back to pre-Greek sources, the ancient Egyptian hieroglyphic symbol for "district" (*hesp*) being a grid (plane rectangular coördinate system). It would therefore be only natural to expect some reference to spatial coördinates in Greek mathematics. But in the whole history of Greek mathematics no such reference is found. Longitude (*mēkos*) and latitude (*platos*) as spherical

[41] *Ibid.*, 557.
[42] Aristotle, *Physics*, 209 a 4–6.
[43] Aristotle, *De caelo*, 268 a 7–10.

coördinates on the celestial sphere or on the earth's surface were
obviously used by Eratosthenes, Hipparchus, Marinus of Tyre,
and Ptolemy, being the ideal two-dimensional system for concentric spheres in Aristotle's world of spherical symmetry. Simplicius
mentions in his commentary on the first book of Aristotle's *De
caelo* that Ptolemy composed an essay *On Extension* (*Peri dia-
staseon*) in which he demonstrated that bodies can have three
dimensions. Moritz Cantor[44] refers to this passage and says: "Bei
der Unbestimmtheit dieser Angabe müssen wir allerdings dahin
gestellt sein lassen, ob man glauben will, es seien in jener Schrift
Gedanken enthalten gewesen, welche dem Begriffe von Raum-
koordinaten nahe kommen." So our assertion of the absence of
spatial coördinates in Greek mathematics may stand. The use of
a three-dimensional coördinate system, and in particular of a
rectangular spatial coördinate system, was not thought reasonable
until the seventeenth century (Descartes, Frans van Schooten,
Lahire, and Jean Bernoulli), when the concept of space had
undergone a radical change. Undoubtedly, Greek mathematics
dealt with three-dimensional objects; Euclid himself, as related
by Proclus,[45] saw perhaps in the construction and investigation
of the Platonic bodies the final aim of his *Elements*. Yet space, as
adopted in mechanics or in astronomy, had never been geome-
trized in Greek science. For how could Euclidean space, with its
homogeneous and infinite lines and planes, possibly fit into the
finite and anisotropic Aristotelian universe?

[44] M. Cantor, *Vorlesungen über Geschichte der Mathematik* (Leipzig,
1880), vol. 1, p. 357.
[45] Procli Diadochi, *In primum Euclidis elementorum librum commentarii*
(Leipzig, 1873), p. 64.

CHAPTER 2

JUDEO-CHRISTIAN IDEAS

ABOUT SPACE

Apart from metaphysics and physics proper, theology proved to be a most important factor in the formulation of physical theories of space from the time of Philo to the Newtonian era and even later. Because of the great effect of theological considerations upon the development of mechanics, as shown in the case of d'Alembert[1] or of Maupertuis[2] who derived from his theoretical physics a proof of the existence of God, it will be worth while looking into the matter. It may be objected to such an inquiry that a religious physicist would quite naturally insist, without recourse to tradition, upon linking science and re-

[1] J. R. d'Alembert, *Oeuvres philosophiques* (Paris, 1805), vol. 2, *Elémens de philosophie,* chap. 6, p. 124.

[2] P. L. M. de Maupertuis, *Essai de cosmologie* (Lyons, 1756).

ligion. Yet it should be borne in mind that the general climate of opinion is historically conditioned.

Let us take for example Colin Maclaurin's *Account of Sir Isaac Newton's philosophical discoveries,* published in London in 1748 by Patrick Murdoch from the author's manuscript papers. In Book One, Chapter One, we read: "But natural philosophy is subservient to purposes of a higher kind, and is chiefly to be valued as it lays a sure foundation for natural religion and moral philosophy; by leading us, in a satisfactory manner, to the knowledge of the Author and Governor of the Universe. To study nature is to search into his workmanship: every new discovery opens to us a new part of his scheme." As this passage shows, the scientist's attitude toward the very function of science may effect his work, while his mental disposition is generally to a great extent determined by his place in history and by his environment. As far as our problem is concerned, there is no doubt that a clearly recognizable and continuous religious tradition exerted a powerful influence on physical theories of space from the first to the eighteenth century.

This influence culminated in the assertion that space is but an attribute of God, or even identical with God. To Newton, absolute space is the sensorium of God; to More, it is divine extension. What are the sources of these doctrines and where did they originate? It is the purpose of this chapter to show that these sources can be traced back to Palestinian Judaism during the Alexandrian period. But this by itself is not enough; we have also to point out the possible channels through which this Eastern lore passed into Western thought.

The earliest indication of a connection between space and God lies in the use of the term "place" (*maḳom*) as a name for God in Palestinian Judaism of the first century. "In Greek philosophy the use of the term 'place' as an appellation of God does not occur." [3] The only intimations of Greek influence in this usage

[3] H. A. Wolfson, *Philo: Foundations of religious philosophy in Judaism, Christianity, and Islam* (Harvard University Press, Cambridge, 1947), vol. 1, p. 247.

are found in Sextus Empiricus and perhaps in Proclus.[4] In Sextus Empiricus we read:

And so far as regards these statements of the Peripatetics, it seems likely that the First God is the place of all things. For according to Aristotle the First God is the limit of Heaven. Either, then, God is something other than the Heaven's limit, or God is just that limit. And if He is other than Heaven's limit, something else will exist outside Heaven, and its limit will be the place of Heaven, and thus the Aristotelians will be granting that Heaven is contained in place; but this they will not tolerate, as they are opposed to both these notions, — both that anything exists outside of Heaven and that Heaven is contained in place. And if God is identical with Heaven's limit, since Heaven's limit is the place of all things within Heaven, God — according to Aristotle — will be the place of all things; and this, too, is itself a thing contrary to sense.[5]

With reference to these words Fabricius adds the following interesting remarks: "Deum Hebraei non dubitant, quia a nullo continetur, ipse vero immensa virtute sua continet omnia, appellare "makom" sive locum, ut saepe fit in libello rituum Paschalium quem edidit Rittangelius." [6]

We have quoted at length the passage from Sextus Empiricus in order to emphasize the fact that for Greek thought the association of God with space is, if admissible at all, only a very remote abstract deduction of an almost paradoxical character; while in Jewish theology of this period, and probably even earlier, the substitution of "place" for the name of God is a common procedure. It seems reasonable to assume that originally the term "place" was used only as an abbreviation for "holy place" (makom kadosh), the place of the "Shekinah." Incidentally, the Arabic term makām designates the place of a saint or of a holy tomb. Since theological conceptions in Judaism soon became more and more abstract and universal, the original connotation of the term "place" fell into oblivion and the word became an

[4] E. Diehl, ed., Procli Diadochi in Platonis Timaeum commentarium (Leipzig, 1903–06), 117 d.

[5] Sextus Empiricus, Adversus mathematicos (Against the physicists), II, 33, trans. by R. G. Bury (Loeb Classical Library; Harvard University Press, Cambridge, 1936), vol. 3, p. 227.

[6] J. A. Fabricius, ed., Sexti Empirici opera (Leipzig, 1840–41), p. 681.

appellation of God without the implication of any spatial limitation. For the notion of God's omnipresence very early became
an important idea, as seen, for example, in the writings of the
Hebrew Psalmist. Thus in Psalm 139 we read:

> Whither shall I go from they spirit? or whither shall I flee from
> thy presence?
> If I ascend up into heaven, thou art there: if I make my bed in
> hell, behold, thou art there.
> If I take the wings of the morning, and dwell in the uttermost
> parts of the sea;
> Even there shall thy hand lead me, and thy right hand shall
> hold me . . .
> My substance was not hid from thee, when I was made in secret,
> and curiously wrought in the lowest parts of the earth.[7]

In the *Midrash Rabbah* we find the following discussion: "R.
Huna said in R. Ammi's name: Why is it that we give a changed
name to the Holy One, blessed be He, and that we call Him
'the place'? Because He is the place of the world. R. Jose b.
Ḥalafta said: We do not know whether God is the place of His
world or whether His world is His place, but from the verse
'Behold, there is a place with me'[8] it follows that the Lord is
the place of His world, but His world is not His place. R. Isaac
said: It is written 'The eternal God is a dwelling-place';[9] now
we do not know whether the Holy One, blessed be He, is the
dwelling-place of His world or whether His world is His dwelling-place. But from the text 'Lord, Thou hast been our dwelling-
place'[10] it follows that the Lord is the dwelling-place of His
world but His world is not His dwelling-place."[11]

The *Mishna* could also be cited as illustrating the frequent use
of the term "place" to denote God.[12] It has been maintained that
this use of the term is of Persian origin, but according to Wolf-

[7] Ps. 139: 7–10, 15.
[8] Exod. 33:21.
[9] Deut. 33:27.
[10] Ps. 90:1.
[11] *Midrash Rabbah, Genesis II, LXVIII,* 9 (trans. by H. Freedman; Son-
cino Press, London, 1939), p. 620.
[12] *Aboth* II, 9; *Pessah* X, 5; *Middoth* V, 4.

son[13] it is undoubtedly of native Jewish origin. Elisaeus Landau[14] referred to certain Pahlavi texts and to the much earlier Zend Avesta in which space is deified and reverenced, as mentioned also by Damascius. Landau traces the use of the term "place" or "space" as an appellation of God back to Simon b. Shetah who, according to *Jerus. Berachoth*[15] had frequent contact with Persians. But Marmorstein[16] has shown that the use of "space" in this sense dates back before Simon b. Shetah, at least to Simon the Just (*c.* 300 B.C.), which disproves Landau's theory of Persian influence; and Geiger[17] disposes of a possible Alexandrian origin.

The Jewish Palestinian metonymy is not to be taken simply as a metaphor, but is evidently the result of a long process of theological thought which culminated in the concept of the Divine Omnipresence. And if this is so, it provides additional support for the assumption of Palestinian origin. For such an ideological development is congenial only to a monotheistic religion and is alien to any form of polytheism whatever. Thus Schechter notes the various steps in the gradual widening of the Divine abode in Jewish theology: "Say the Rabbis, 'Moses made Him fill all the space of the Universe, as it is said "The Lord he is God in the heaven above, and upon the earth beneath; there is none else," [18] which means that even the empty space is full of God." [19]

In the Palestinian Talmud we find an interesting legend that admirably illustrates our thesis: R. Tanhuma narrates how a horrible storm threatened a boat on which a company of pagans and one Jewish boy were sailing; as their life seemed to be in danger, each passenger reached for his idol to worship it, but without success; finally the Jew yielded to the pagans' request

[13] Wolfson, *Philo*, vol. 1, p. 247.
[14] E. Landau, *Die dem Raum entnommenen Synonyma für Gott* (Zurich, 1888), p. 42.
[15] *Palestinian Talmud, Berachoth*, VII, 2, *Nasir* V, 5.
[16] A. Marmorstein, *The old rabbinic doctrine of God* (London, 1927), vol. 1, p. 92.
[17] A. Geiger, *Nachgelassene Schriften* (Breslau, 1885), vol. 4, p. 424.
[18] Deut. 4:39.
[19] S. Schechter, *Some aspects of rabbinic theology* (New York, 1910), p. 25.

and prayed to his God, whereupon the sea calmed down. When they came to the nearest port, all of them went ashore to buy provisions, except the Jew. When asked why he stayed aboard, he replied, "What can a needy stranger like me do?" Whereupon they answered, "You a poor stranger? We are the strangers. We are here, but our gods are in Babylon or Rome; and others amongst us who carry their gods with them derive not the least benefit from them. But you, wherever you go, your God is with you." [20]

This notion, that God was at one and the same time here and there, had no pantheistic consequences in Jewish theology, but led to the association of God and space as an expression of his ubiquity. This usage had spread to Alexandrian philosophy,[21] was incorporated in the Septuaginta[22] and was adopted also in the pre-Mohammedan world of thought, as we see from the *Divan* of Lebid.[23] In later Jewish literature the term "space" or "place" as a name for God became so frequent that an explanation, if only *post facto*, appeared to be called for. Thus the gematria explains that both the name of God (*the nomen ineffabile*) and the word "place" lead to the same number: adding the squares of the numbers corresponding to the letters of the holy name one gets the sum of the numbers that correspond to the letters of the word "place":

$$40 = \text{ם} \quad 6 = \text{ו} \quad 100 = \text{ק} \quad 40 = \text{מ}, \qquad 5 = \text{ה} \quad 6 = \text{ו} \quad 5 = \text{ה} \quad 10 = \text{י};$$
$$40 + 6 + 100 + 40 = 186, \qquad 5^2 + 6^2 + 5^2 + 10^2 = 186.$$

The designation "place" for God and the mystical conception of God as the space of the universe are frequently encountered in the post-Talmudic-Midrashic literature. The *Zohar* vindicates this use by saying that God is called "space" because He is the space of Himself.[24] As is well known, the *Zohar* is a collection

[20] *Berachoth*, IX, Halacha 1.
[21] See Philo, *De somniis*, I, 575.
[22] Compare Exod. 24:10 with the Hebrew original.
[23] Jūsuf Dijā-ad-Dīn al-Chālidī, ed., *Der Diwan des Lebid* (Vienna, 1880), p. 12.
[24] *Zohar*, I, 147 b; II, 63 b and 207 a.

of treatises, texts, and extracts belonging to different periods, but with one common purpose: to reveal the hidden truth in the Pentateuch. Tradition claims that Simon b. Yohai, a sage of the second century and pupil of Rabbi Akiba, was the author of this book of mystical interpretation. According to the legend, he had spent many years in complete solitude, when he received sacred revelations from the prophet Elijah. It is said that the *Zohar* has been hidden for more than a thousand years in a cave in Galilee until it was discovered by Moses of Leon at the end of the thirteenth century. According to another version, Moses of Leon compiled the *Zohar* himself, using an Aramaic idiom to give it an air of antiquity. Whatever its source may be, it is a collection of ancient Jewish folklore and oral tradition that had a great influence, and not on Jewish thought alone. Italy especially, that shifting conglomeration of republics, states, and cities in the Renaissance, became a fruitful soil for Jewish esoteric teachings and, in particular, for the spread of cabalistic ideas. In the second half of the fifteenth century, when Greek scholars streamed westward after the fall of Constantinople in 1453, there were among them Jewish scholars who found a refuge in Italy, as we know from the case of Elijah del Medigo. In 1480 Elijah was called to the University of Padua, where he made the acquaintance of Giovanni della Mirandola, who invited him to Florence. Pico della Mirandola is generally considered as the first to introduce the cabala into Christianity. By the time of Mirandola's death, in the year 1494, cabalistic ideas had already spread farther to the north. Henry Cornelius Agrippa of Nettesheim, who became a lifelong devotee of the cabala, delivered a lecture in 1509 at the University of Dôle on Reuchlin's *De verbo mirifico* in which he preached the doctrine of the cabala and which led to a controversy with a Franciscan, who accused him of being a "Judaizing heretic." The promulgation of the *Zohar* toward the end of the sixteenth century in Italy added new impetus to the spreading of cabalistic ideologies to the north, and the number of scholars interested in this rabbinical learning increased yearly.

One of the most erudite scholars of his time, John Rainoldes (1549–1607), president of Corpus Christi College in Oxford, made an extensive study of rabbinical lore. The German alchemist Michael Maier (born in Ruidsburg, Holstein, 1568), who became court physician to the Emperor Rudolph, visited England in 1615, where he very likely exerted great influence on Robert Fludd. Fludd was one of the early English Platonists whose importance for English theories of space and time in the sixteenth century cannot be overlooked. He taught "the immediate presence of God in all nature" and illustrated his ideas from Trismegistus. "God is the center of everything, whose circumference is nowhere to be found." One of the major sources of Fludd's rabbinical knowledge was probably Rainoldes' *Censura librorum apocryphorum Veteris Testamentis*.[25]

But before we continue with the history of the English Christian Cabalists and Platonists and their influence on theories of space and time, let us return to Italy, where Campanella, a leading figure in the new Italian natural philosophy, was engaged in formulating a spiritualized conception of space.

If we bear in mind that both Newton's and Locke's theories of space originated in part, at least, in Gassendi's natural philosophy and that Gassendi was in personal contact with Campanella,[26] we are in a position to see that Campanella's ideas about space are of no little importance for the history of later natural philosophy. As can be shown in detail by comparing the writings of Campanella, especially his astrological and metaphysical works, or his *Medicinalium*," [27] with the writings of Paracelsus and of Agrippa of Nettesheim, the influence on Campanella of German mystical thought of the middle of the sixteenth century was very great. But, as we know, Paracelsus, and Agrippa much more, were ardent students of the Jewish cabala. It is no wonder,

[25] Oppenheim, 1611.
[26] They met, for example, in Aix. See Gassendi's biography of Peiresc.
[27] Thomas Campanella, *Medicinalium juxta propria principia libri septem* (Lyons, 1635).

therefore, that Campanella's ways of thinking show strong cabalistic tendencies. In his *De occulta philosophia* (liber I) Agrippa
restates in the spirit of the cabala the doctrine of the realization
of the Divine Thought through the creation of a hierarchy of
worlds. With Campanella the hierarchy of five mutually envelop-
ing and penetrating worlds comprises the *mundus mathematicus
seu spatium*, the third realization after the *mundus Archetypus*
and the *mundus mentalis*. In his *Metaphysicarum rerum juxta
propria dogmata*[28] Campanella characterizes this mathematical
world or space as the "omnium divinitas substentas, portansque
omnia verbo virtutis suae . . ." In words analogous to those of
R. Huna,[29] "The Lord is the dwelling-place of His world but
His world is not His dwelling-place," Campanella states that
space is in God, but God is not limited by space, which is His
"divina creatura."[30] The idea of the identification of space with
at least an attribute of the Divine Being gains new impetus in
the second book,[31] where we read: "Spatium, entia locans invenio
primum immortale, quia nulli est contrarium." In Campanella's
conception, space becomes an absolute, almost spiritual entity,
characterized by divine attributes. Its reality guarantees a sound
foundation for mathematical speculations, which, according to
Campanella, ought to be based not on hypothetical artifacts but
on reliable sense data.

Together with this cabalistic-Platonic conception we find in
Campanella's theory of space the strong influence of his teacher
Bernardino Telesio. We shall have occasion in the next chapter
to speak of Telesio since his doctrine represents a turning point
in the history of physical thought of the sixteenth century owing
to its anti-Aristotelian conception of space and time. But in the
frame of the present chapter we are concerned only with the

[28] Paris, 1638.
[29] See p. 28.
[30] Campanella, *De sensu rerum et magia*, libri quatuor (Frankfurt, 1620),
I, c. 12.
[31] *Ibid.*, c. 26.

change brought about in Campanella's original conception of space by Telesio's critique. Campanella, too, came to the conclusion that space was completely homogeneous and undifferentiated, immovable and incorporeal, penetrated by matter and penetrating matter, destined for the collocation of mobile entities. "Up" and "down," "right" and "left" were only pure creations of the intellect, designed to facilitate practical orientation among the multitude of concrete bodies, but with no real directional differentiations in space corresponding to them. Campanella is here following in the footsteps of his master, as he generally did, and to such an extent that he was held to be the reincarnation of him: ". . . horum clarissimus erat Thomas Campanella Stylensis, cujus in corpus Telesii ingenium transmigrasse dicebatur." [32]

Another trend in the history of the theory of space, very similar in its mystical-theological character and its association of God and space, was the identification of space and light. From prehistoric times light was the symbol of supernatural forces. The oldest religions are of astral character, as Egypt's Re and ancient Persia's Ahura Mazda bear witness. In the *Sutran*[33] Brahman is personified as the Primordial Light. Atman is honored by the Gods "as immortal life, as the light of lights." [34]

Even the Bible, which prohibits all images of God, still uses the element of light as the medium in which God becomes visible to man: God appeared to Moses in a burning bush;[35] a column of fire showed the children of Israel the way out of Egypt.[36] We read in the Psalms: "Who coverest thyself with light as with a garment." [37] In the New Testament God is explicitly identified

[32] *Erythraeus,* Pinacotheca, I, p. 41.
[33] *Sutran,* I, 3. 22–23.
[34] *Brih ad-aranyakam,* 4, 4, 16; cf. P. Deussen, *The system of the Vedānta* (Chicago, 1912), p. 130.
[35] Exod. 3:4.
[36] Exod. 13:21; Num. 14:14.
[37] Ps. 104:2.

with light: "Ego sum lux mundi." [38] Light plays an important role in the metaphysical systems of Philo and of Plotin. In the vocabulary of the Jewish Midrashim and of the cabala "light" is one of the most important terms signifying the most holy conception. The word *Zohar*,[39] the title of the most important book in Jewish mysticism, denotes "light," "splendor," or "glimmer." According to the cabala, the Infinite Holy One, whose light originally occupied the whole universe, withdrew his light and concentrated it on his own substance, thereby creating empty space.[40]

This apotheosis of light became a fundamental characteristic of later Neoplatonism and medieval mysticism. Even the more sober natural philosophy of the Middle Ages, though still anthropomorphic with its hierarchy of values in nature, accepted light as the most "noble" entity in the world. Plotinus set the example of ranking light as highest in existence. Through its various degrees and emanations the macrocosm formed one coherent and organic unit. Light is the means by which universal order is maintained. In its purest actuality light is Deity. According to Saint Bonaventure, God is "spiritualis lux in omnia-moda actualitate." [41] The theories that identify space with light under the influence of Neoplatonism and religious mysticism are therefore essentially theological in character. In our treatment of them here we shall confine ourselves in the main to two representative examples, one in antiquity, the other in the thirteenth century. The first is the theory of Proclus; the second is that of Witelo.

Fragment No. 6 of *The fragments that remain of the lost writings of Proclus*[42] discusses space as being the interval between the boundaries of the surrounding body. This is similar to the

[38] John 8:12.
[39] The name is derived from Dan. 12:3.
[40] The cosmogony of Genesis does not conceive space as a product of creation. But God's first words were: "Let there be light" (Genesis 1:3). Thus light was created before the stars or the sun existed.
[41] *Liber sententiarum*, II, 13 a. Cf. Etienne Gilson, *La philosophie de Saint Bonaventura* (Vrin, Paris, 1943), p. 217.
[42] English translation by Thomas Taylor (London, 1825), p. 113.

Stoic definition. But Proclus draws the conclusion that space must
therefore be commensurable with things that are corporeal and
since it is able to contain other bodies it must be in some sense
corporeal in itself. Yet space must be immaterial and immovable;
for were it material, it would not be able to occupy the same
place as another body, and were it movable, it would be moving
in space, that is, in itself. To Proclus, space contains the whole
material world, but is not contained by the world and becomes
therefore coextensive with the domain of light.[43] It is regrettable
that Proclus's work *On space*, mentioned in various texts,[44] is not
extant; thus we know very little about the immediate source
of this identification of space and light. Perhaps it was an old
Pythagorean theory about the primitive light or the more poetical
variation in Plato's *Republic* where the myth of the Pampylian
Er, son of Armenios, is related: "For this light binds the sky together, like the hawser that strengthens a trireme, and thus holds
together the whole revolving universe." [45]

The conceptions of the Neoplatonic "light-metaphysics," propagated persistently also by Jewish philosophy and mysticism
(Saadya of Fayum, Ibn Gabirol and many cabalists), exerted a
strong influence on Robert Grosseteste. Assuming light (*lux*) as
the first corporeal form and the first principle of motion, Grosseteste reduced the creation of the universe in space to the "auto-
diffusion" of light. Light, which in his view is propagated by
itself instantaneously, as shown by its physical manifestation,
the visible light, is conceived by him as the basis of extension
in space. As A. C. Crombie rightly points out in his book on
Grosseteste,[46] it was this fundamental assumption that made
Grosseteste believe that the key to the understanding of the
universe lies in the study of geometric optics. Thus, in the last

[43] Simplicius, *Physics*, 612, 32.
[44] For example, in the *Fihrist*.
[45] Plato, *Republic*, X, 616.
[46] A. C. Crombie, *Robert Grosseteste and the origins of experimental
science* (Clarendon, Oxford, 1953), p. 104.

analysis, it was the Neo-Platonic conception of space that led to the great interest in optics and mathematics exhibited by the thirteenth century.

Among the most conspicuous figures in this respect were the unknown author of the *Liber de intelligentiis* and the Silesian Witelo, both of whom were certainly influenced by Grosseteste.[47] In fact, the writings of these two are so similar both in contents and in style that Clemens Baeumker erroneously attributed the *Liber de intelligentiis* to Witelo.[48] As with Proclus, the point of departure in the unknown author's theory is Aristotle's physics. We read at the very beginning of the *Liber de intelligentiis* that space is the "ultimum continentis immobilis," in contrast to Aristotle's "ultimum immobile continentis." As Baeumker remarks, the author accepted the introduction of two additional celestial spheres, the second of which is immovable, the motive being to bring science into line with Scripture.[49] But when the author comes to the famous passage in Aristotle's *Physics* reading, "Moreover the trends of the physical elements (fire, earth, and the rest) show not only that locality or place is a reality but also that it exerts an active influence," [50] he becomes Platonic in his conception of "dynamis" (power, in the sense of exertion of force) and interprets it as the faculty of space. In his view, the nature of space is characterized by two functions: to inclose (the *periechein* of Aristotle) and to support "continere" and "conservare." Light, the source of all existence, the all-pervading power, ranking highest in the hierarchy of Being, light alone fulfills these two conditions. Hence space and light are one.

[47] A. Birkenmajer, "Études sur Witelo," *Bull. intern. acad. polon., Classe hist. philos.* (Cracow), p. 4 (1918), p. 354 (1920), p. 6 (1922).
[48] C. Baeumker, "Zur Frage nach Abfassungszeit und Verfasser des irrtümlich Witelo zugeschriebenen Liber de intelligentiis," in *Miscellanea Francesco Ehrle* (Studi e testi, 37–47) (Biblioteca Apostolica Vaticana, Vatican City, 1924). The *Liber de intelligentiis* was edited with a commentary by Baeumker (Münster, 1908).
[49] Cf. the "caelum Empyreum" of William of Auvergne.
[50] Aristotle, *Physics*, 208 b 10–11.

The assertion, "Unumquodque primum corporum est locus et forma inferiori sub ipso per naturam lucis," [51] is proved by a series of syllogisms. [52]

These writers were not the last proponents of the importance of light for a theory of space. Light played an important role in most cosmologies of the Italian natural philosophers of the Renaissance. Francesco Patrizi, the predecessor of Campanella, was also fascinated by the mysteries of light and made light an integral part in his speculations. He was faced, like most philosophers of nature, in the sixteenth century, with the formidable task of incorporating the inherited supernatural world of the Middle Ages in the newly discovered world of nature of the Renaissance. The problem was how to unite the corporeal concrete world of nature with the incorporeal world of spirit. Space, light, and soul, and the Neoplatonic doctrine of emanations, together constitute the means by which he tried to solve the problem. Space, the entity that is neither corporeal nor immaterial, serves as the intermediary between the two worlds. Indeed, it was created by God for the fulfillment of this function. Space is infinite, for an infinite cause can give rise only to an infinite effect. The traditional scholastic principle, according to which the effect is always feebler than the cause, applies, according to Patrizi, only to finite entities. The first thing to fill this space is light, the all-pervading, all-preserving medium of three dimensions, whose importance is not confined to its physical function as a transmitter of heat, power, and other influences; it is also metaphysically the way to God.

An outstanding example of a strong religious bias in the con-

[51] *Liber, de intelligentiis,* VIII, 4. (See reference 48.)

[52] "Cuius expositio est quod locus est ultimum continentis immobilis; illud autem ultimum caeli est ultimum per comparationem ad id ad quod determinatur locus unicuique inferiori sub ipso, sicut manifestum est de naui et palo fixo in aqua. mutat enim superficiem corporis continentis, scilicet aquae, non tamen mutat locum quia caeli non mutat partem, per comparationem ad quam determinabatur ei locus. unde caeli ultimum locus est. Hoc autem habet naturam lucis. Illud enim ultimum est continens et conseruans, cum sit locus . . ."

ception of space is the theory of Henry More.[53] In order to sub-
stantiate his vigorous religious beliefs, More thought it necessary
to augment the science of Descartes with cabalistic and Platonic
concepts. That the cabala, apart from Neo-Platonic philosophy,
was a major factor in More's conception of space can be proved
not only by an analysis of the conception itself, but on historical
grounds as well. First of all, we know that More, together with
Fludd, was regarded as one of the greatest rabbinical students of
his time. He certainly studied Hebrew and read the Scripture
in its original version, as may be seen from his wide use of that
language in his various writings, especially in his *Discourses on
several texts of Scripture*.[54] Having studied the writings of Mar-
silius Ficinus, Plotinus, and Trismegistus, he was persuaded that
cabalistic philosophy, as expounded, for instance, in the *Liber
drushim* (Book of dissertations) of Isaac Luria, was of the greatest importance. As far as his theory of space is concerned, More
himself refers to the cabalistic doctrine as explained by Cornelius
Agrippa in his *De occulta philosophia*,[55] where space is specified
as one of the attributes of God.

More's metaphysical writings are in the main a somewhat
desultory expansion of his fundamental views on the nature of
incorporeal substances, whose existence he was convinced that
he had proved on the basis of his cabalistic studies. During the
last twenty years of his life he wrote numerous publications on
mystical subjects, including a *Cabalistic catechism*.[56] Some of
these writings were addressed to Baron Knorr von Rosenroth and
published in the *Kabbala denudata*,[57] a translation into Latin of

[53] For his biography and intellectual development, see R. Ward, *The life
of Dr. H. More* (London, 1710).
[54] "By the late pious and learned Henry More, D. D." (London, 1692).
[55] III, 11. Cf. Friedrich Barth, *Die Cabbala des H. C. Agrippa von Net-
tersheim* (Stuttgart, 1855).
[56] For a list of More's cabalistic writings, see Gerhard Scholem, *Biblio-
graphia kabbalistica* (Berlin, 1933), p. 110.
[57] *Kabbala denudata seu doctrina Hebraeorum transcendentalis et metaphysica* (Sulzbach, 1677).

cabalistic writings that has done an immense service to many occultists by furnishing material for their reveries.[58]

More exerted a great influence on Locke, Newton, and Clarke, and through them on eighteenth-century philosophy in general, so that his doctrine is worth analyzing in detail. His writings reveal that the problem of space occupied his mind as early as 1648 at least and continued to interest him till 1684; that is, his interest dates from his correspondence with Descartes to his correspondence with John Norris. In his letters to Descartes More places himself in strong opposition both to ancient Greek atomism and to the Cartesian identification of space and matter.. Both seem to him to lead inevitably to materialism and atheism, and for the refutation of these he deems it necessary to demonstrate the existence of a spiritual Being, which permeates and acts in nature. To Descartes the attribute that distinguishes spirit from matter is thought, as manifested in contemplation and consciousness; to More it is spontaneous activity, the source of all change and motion. Since change and motion are present in all realms of nature, the question arises how this interaction can be performed on matter at all. The answer lies, according to More, in the nature of space, the clear understanding of which can alone save philosophy from an otherwise inevitable atheism. "Atque ita per eam ipsam januam per quam Philosophia Cartesiana Deum videtur velle e Mundo excludere, ego, e contra, eum introducere rursus enitor et contendo." [59]

The chief motive behind More's concern with the problem of space, as it is the motive of his whole philosophy, is to find a convincing demonstration of the indubitable reality of God, spirit, and soul. In accord with this purpose he rejects Descartes' absolute identification of matter and extension. In order to prove the reality of spirit, it suffices to show that extension is

---

[58] On the relations between More, Knorr, and Van Helmont, see John Tulloch, *Rational theology and christian philosophy in England in the 17th century* (Edinburgh and London, 1872), vol. 2, p. 345.

[59] Henry More, *Enchiridion metaphysicum sive de rebus incorporeis* (London, 1671), part I, chap. 8, 7.

spiritual provided extension itself is real. On the basis of this reasoning More's treatment of the space problem may be divided into three parts: (1) extension is not the distinguishing attribute of matter; (2) space is real, having real attributes; (3) space is of divine character. We shall comment on each of these in turn.

(1) In his correspondence with Descartes More points out that, apart from the primary qualities of matter, mentioned by Descartes, matter has also the property of impenetrability or "solidity," as impenetrability was called at that time. Impenetrability (and the related tangibility) is the *criterium differentionis* between matter and extension. As is well known, Locke in his *Essay concerning human understanding*[60] takes account of these ideas. In order to understand the mutual interaction between the world of spirit and the world of matter it is necessary to find a common ground between them. This common ground is space. Hence it follows that extension characterizes the world of spirit no less than the world of matter. In a word, extension is not a distinguishing attribute of matter, but belongs to both spirit and matter.

(2) In order to prove that space is real, More used different arguments at different times. Already in his correspondence with Descartes he undertakes to refute the latter's plenum against the existence of space as such. To both thinkers empty space does not exist; but if space may be empty as far as matter is concerned, it is yet, according to More's view, always filled with spirit. Descartes contended that the walls of a vessel that is exhausted of air must necessarily collapse. "Si quaeratur, quid fiet, si Deus auferat omne corpus quod in aliquo vase continetur, et nullum aliud in ablati locum venire permittat? Respondendum est: Vasis latera sibi invicem hoc ipso fore contigua." [61] Descartes' prediction of this presumably too complicated experiment, which in his view only a God is able to perform, was based on his philosophy of space and matter. Descartes published his *Principia*

[60] Book II, c. 4.
[61] Descartes, *Principia philosophiae*, II, 18.

in 1644. Only a few years later, or perhaps even at the same time,[62] a simple burgomaster performed such an experiment. *Sed vasis latera non fierunt contigua!* When reading today Descartes' argumentation on this subject one has to bear in mind that the concept of "a sea of air," surrounding the earth, evolved only toward the middle of the seventeenth century, as the history of pneumatics shows, and that it was unknown to the French philosopher.[63] More's reply to Descartes' argument must be understood with these same cautions in mind: It is not necessary that the walls of the vessel collapse, since all motion of matter in Descartes' own view is originated in God, so that God may impart to the walls of the vessel some counteracting motion and so prevent its collapse.[64] Yet even if God can provide for the existence of an empty space, it still would not be an absolute vacuum because of the "divine extension" which permeates all space.

Still another proof for the reality of space, more scholastic in manner, is given in an appendix to More's *Antidote against atheism*.[65] The existence of space is guaranteed by its very measurability "par aunes ou par lieues." [66] In other words, space has indubitably the attribute of measurability, even when empty of all matter; as there are no accidents without substance, measurability as an accident demonstrates the substantiality of space. It is of course an incorporeal substance, since it includes "certain notions, such as immobility and penetrability, which are inconsistent with matter." [67] The penetrability of space affords

[62] We do not know the exact date when Otto von Guericke began his famous line of experiments. It was between 1635 and 1654. In 1657 the first accounts of the air-pump experiments were published by Kaspar Schott in his *Mechanica hydraulico-pneumatica,* through which Robert Boyle became acquainted with Guericke's experiments.
[63] See J. B. Conant, *On understanding science* (Mentor, New York, 1952), p. 54.
[64] *Oeuvres de Descartes* (Paris, 1824–1826), vol. 10, p. 184.
[65] First edition published 1653 (appendix 1655, London).
[66] *Oeuvres de Descartes,* vol. 10, p. 214.
[67] More, *Enchiridion metaphysicum,* VI–VIII; letter to Descartes, March 1649.

in More's view another proof of its incorporeality, and hence of its total discriminability from matter. More differs from Descartes, whose doctrine in this respect he characterizes in the following words: "For though it is wittily supported by him for a ground of more certain and mathematical after-deductions in his philosophy; yet it is not at all proved, that matter and extension are reciprocally the same, as well every extended thing matter as all matter extended. This is but an upstart conceit of the present age." [68]

In the book just quoted, another very interesting demonstration of the reality of space is attempted, which, because of its originality and the probability of its direct influence upon Newton, calls for detailed treatment. Philotheus, the "zealous and sincere Lover of God and Christ, and of the whole Creation," in his discourse with Hylobares, the "young, witty, and well-moralized Materialist," gives the following example:

Philotheus: Shoot up an Arrow perpendicular from the Earth; the Arrow you know, will return to your foot again.

Hylobares: If the wind hinder not. But what does this Arrow aim at?

Philotheus: This Arrow has described only right Lines with its point, upwards and downwards in the Air; but yet, holding the motion of the Earth, it must also have described in some sense a circular or curvilinear line.

Hylobares: It must be so.

Philotheus: But if you be so impatient of the heat abroad, neither your body nor your phancy need step out of this cool Bowre. Consider the round Trencher that Glass stands upon; it is a kind of a short Cylinder, which you may easily imagine a foot longer, if you will.

Hylobares: Very easily, Philotheus.

Philotheus: And as easily phansy a Line drawn from the top of the Axis of that Cylinder to the Peripherie of the Basis.

Hylobares: Every jot as easily.

Philotheus: Now imagine this Cylinder turned round on its Axis. Does not the Line from the top of the Axis to the Peripherie of the Basis necessarily describe a Conicum in one Circumvolution?

Hylobares: It does so, Philotheus.

Philotheus: But it describes no such Figure in the wooden Cylinder itself: As the Arrow in the aiereal or material Aequinoctial Circle

[68] More, *Divine dialogues* (London, 1668), I:XXIV.

describes not any Line but a right one. In what therefore does the one describe, suppose, a circular Line, the other a Conicum?

Hylobares: As I live, Philotheus, I am struck as it were with Lightning from this surprizing consideration.

Philotheus: I hope, Hylobares, you are pierced with some measure of Illumination.

Hylobares: I am so.

Philotheus: And that you are convinced, that whether you live or no, that there ever was, is, and ever will be an immovable Extension distinct from that of movable Matter.[69]

In these last words, reminiscent of the style of Jewish medieval Piutim, Philotheus concludes with the absolute reality of space. It is interesting to note that in the case of the arrow, the proof is based, in the last analysis, on the relative motion of the earth according to the Copernican theory, and in the case of the rotating cylinder the proof is based on the assumption that rotation is always relative to something, a question that occupied Newton, as we know from his famous experiment with the rotating pail. Whereas in the first case no observable phenomena result, the second, according to Newton, makes the existence of absolute space physically demonstrable.

Space, then, is the medium in which the curved line or the cone is formed. But here the discussion is interrupted by Cuphophron, "a zealous, but Aiery-minded, Platonist and Cartesian, or Mechanist," who contends, "that it may be reasonably suggested, that it is real Extension and Matter that are terms convertible; but that Extension wherein the Arrow-head describes a curvilinear Line is only imaginary." This remark ushers in a line of arguments that for the modern reader suggests the Kantian conception of space. For Hylobares replies: "But it is so imaginary, that it cannot possibly be dis-imagined by humane understanding. Which methinks should be no small earnest that there is more than an imaginary Being there."

Collateral confirmation of the reality of space is adduced by calling upon the authority of the ancient atomists, of Aristotle, and of the Pythagoreans with reference to the famous analogy

[69] *Ibid.* (London, ed. 2, 1713), p. 52.

that the "Vacuum were that to the Universe which the Air is to particular Animals." Finally, More reverts to his old argument, which we characterized as scholastic, when he makes Hylobares say: "And lastly, O Cuphophron, unless you will flinch from the Dictates of your so highly-admired DesCartes, forasmuch as this Vacuum is extended, and measurable, and the like, it must be a Reality; because Non entis nulla est Affectio, according to the Reasonings of your beloved Master. From whence it seems evident that there is an extended Substance far more subtile than Body, that pervades the whole Matter of the Universe."

(3) It is contended with regard to this "subtile" substance, called further on the "Divine Amplitude," that it exists necessarily, and would exist even if all matter were annihilated. The necessary existence of space, even without matter, leads More to the final identification of space with God. For it might be argued, as Cuphophron argues toward the end of the discussion on the nature of space, that, if God as well as matter were annihilated from the world, extension would seem necessarily to remain. The spokesman for More's reasoning here is Bathynous, "the deeply-thoughtful and profoundly-thinking Man," who replies that God's essence implies existence (the ontological proof). In other words, to assume the annihilation of God is a *contradictio in adjecto*. God and space have both the property of necessary existence; they are therefore one and the same.

This conclusion that God and space are one is drawn also in the appendix to the *Antidote against atheism:*

If after the removal of corporeal matter out of the world, there will be still space and distance, in which this very matter, while it was there, was also conceived to lie, and this distant space cannot but be conceived to be something, and yet not corporeal, because neither impenetrable nor tangible, it must of necessity be a substance incorporeal, necessarily and eternally existent of itself; which the clearer idea of a being absolutely perfect will more fully and punctually inform us to be the self-subsisting God.[70]

[70] *A collection of several philosophical writings of Dr. Henry More* (London, ed. 2, 1655), appendix, p. 338. Cf. F. I. MacKinnon, ed., *Philosophical writings of Henry More* (London and New York, 1925).

The attributes of space are attributes of God. A list of these attributes is given in More's *Enchiridion metaphysicum:*

Neque enim Reale duntaxat, (quod ultimo loco notabimus) sed Divinum quiddam videbitur hoc Extensum infinitum ac immobile, (quod tam certo in rerum natura deprehenditur) postquam Divina illa Nomina vel Titulos qui examussim ipsi congruunt enumeravimus, qui & ulteriorem fidem facient illud non posse esse nihil, utpote cui tot tamque praeclara Attributa competunt. Cujusmodi sunt quae sequuntur, quaeque Metaphysici Primo Enti speciatim attribunt. Ut Unum, Simplex, Immobile, Aeternum, Completum, Independens, A se existens, Per se subsistens, Incorruptibile, Necessarium, Immensum, Increatum, Incircumscriptum, Incomprehensibile, Omnipraesens, Incorporeum, Omnia permeans & complectans, Ens per Essentiam, Ens actu, Purus Actus. Non pauciores quam viginti Tituli sunt quibus insigniri solet Divinum Numen, qui infinito huic Loco interno, quem in rerum natura esse demonstravimus, aptissime conveniunt: ut omittam ipsam Divinum Numen apud Cabbalistes appellari "maḳom," id est, Locum.[71]

This list of attributes and names, a recurrent theme in the cabalistic writings, is quoted in full to show the extent to which More was influenced by Jewish mysticism. In his *Divine dialogues* as well More mentions the cabalists in connection with the divine nature of God. The discussion that we cited from the *Dialogues* ends with Psalm 90:

Lord, thou hast been our dwelling-place in all generations.
Before the mountains were brought forth, or ever thou hadst formed the Earth or the World, even from everlasting to everlasting, thou art God.

No doubt, the general tenor of the cabala may easily have provoked such spiritual ideas about space as were harbored by More. Anyone who has read the *Book of formation* (*Sepher Yezirah*), which deals with cosmogonical problems of the universe, or who has read on Luria's cabalistic notion of the "Zimzum," [72] the divine self-concentration, creating space by self-restriction, will certainly accept the thesis that a somehow

[71] More, *Enchiridion metaphysicum*, part I, chap. 8.
[72] "Deus creaturus mundos contraxit praesentiam suam," *Kabbala denudata* (Sulzbach, 1677), part II, p. 150.

pantheistic interpretation of the cabala must necessarily lead to More's conception of space.

Indeed, a similar intellectual process was most probably influential on the philosophy of Spinoza. With reference to his fundamental dictum; "Quidquid est, in Deo est et nihil sine Deo esse neque concipi potest," [73] Spinoza admits in a letter to Oldenburg: ". . . omnia, inquam, in Deo esse et in Deo moveri cum Paulo affirmo . . . et auderem etiam dicere, cum antiquis omnibus Hebraeis, quantum ex quibusdam traditionibus, tametsi multis modis adulteratis conjicere licet." [74] As A. Franck in La cabbale[75] and much earlier (1699) Johann Wachter in Der Spinozismus in Jüdenthumb[76] have shown, Spinoza's remarks can refer only to cabalistic writings.

In contrast to More, Spinoza includes not only extension but also matter as an attribute of God, thus changing the conception of God into that of an absolutely impersonal, almost mechanical God, as is shown in his Ethics. Leibniz, who seems to have read Wachter's De recondita Hebraeorum philosophia,[77] in which the author repeats his thesis concerning Spinoza's dependence on the cabala, wrote in this context in a letter to Bourget, "Verissimum est, Spinozam Cabbala Hebraeorum esse abusum." [78]

This digression on Spinoza's philosophy and its possible cabalistic sources has been inserted only to show that elements of Jewish esoteric writings, perhaps owing to the rise of Neoplatonic ideas, could have been easily integrated into the philosophy of the seventeenth century. We reserve judgment

[73] "Ethica more geometrico demonstrata," I, prop. 15, in Spinoza, Opera (ed. C. Gebhart; Heidelberg, 1925), p. 56.

[74] H. Ginsberg, ed., Der Briefwechsel des Spinoza im Urtext (Leipzig, 1876), p. 53, Epistola XXI; also Epistola LXXXIII.

[75] Adolphe Franck, La cabbale ou la philosophie des Hébreux (Paris, 1843).

[76] J. G. Wachter, Der Spinozismus in Jüdenthumb oder die von dem heutigen Jüdenthumb und dessen geheimen kabbala vergötterte Welt (Amsterdam, 1699).

[77] Published in 1706.

[78] Letter of the year 1707.

on the question how far in detail Spinoza, More, or any other thinker of that time has actually been influenced by the cabala, but we claim — as definitely demonstrated in the case of the concept of space — that certain general ideas of cabalistic origin have been absorbed into the intellectual climate of that period.

Our investigation into the influence of Judeo-Christian religious speculations on the conception of absolute space in the seventeenth century has not been presented as an uninterrupted chain of incontestable conclusions. Owing to the evasive character of the rather obscure and mystical ideas involved, the contention is rather based on the tradition of a certain "climate of opinion" or attitude of mind, and not on a direct communication of definite statements. In the case under discussion it was at least possible to expose all the major stages in this transmission.

More problematic, and still more conjectural, is the theory of a religious background in the rival conception of space, that is, Leibniz's relational point of view. As we shall see in Chapter IV. Leibniz rejected Newton's theory of an absolute space on the ground that space is nothing but a network of relations among coexisting things. In his correspondence with Clarke, Leibniz likens space to a system of genealogical lines, a "tree of genealogy," or pedigree, in which a place is assigned to every person. The assumption of an absolute space, according to Leibniz's view, is wholly analogous to a hypostatization of such a system of genealogical relations.

Now it is important to recall that a similar theory had already been propounded by the eleventh-century Muslim philosopher Al-Ghazālī, or possibly by one of his predecessors. Here again it becomes apparent that theological and metaphysical speculations were influential on the formulation of a theory of space. In fact, the whole issue in question is based on the ideological conflict between Aristotelian cosmology and the Koranic dogma of divine creation. For a full understanding of the situation it is perhaps most instructive to discuss both the temporal and the spatial aspect of the problem. Time was defined by Aristotle[79] as the "number of motion" (for example, of the revolutions of the celestial spheres). Since without natural bodies there cannot be motion, Aristotle concluded that outside the finite heaven there is no time. Place, according to the Stagirite, presupposes the possibility of the presence of bodies.[80] Since outside the finite heaven no body can exist, as proved in his writings previously, Aristotle deduced that outside the finite heaven there is no place. So far Muslim philosophy complies with Aristotelian cosmology. But now the conflict becomes apparent: whereas Aristotelian cosmology assumes the eternity of substance, the Koranic dogma affirms divine creation. Thus for Muslim philosophy a new problem is raised which did not exist for Aristotelian cosmology. It is the question whether there were space and time prior to the act of creation. Obviously, the answer must be negative, since space and time have no existence apart from matter *per definitionem;* they are mere relations among bodies. Space and time are consequently also products of creation. Muslim philosophy even went so far that it rejected the logical validity of the statement, "God was before the world was" [81] (*"Kāna allāhu walā 'ālama"*), if *kāna* is to be understood in a temporal sense. Even in the most fundamental proposition, "God created the world" (*"halaka allāhu al 'ālama"*), the verb "created" (*halaka*) has to be understood in a causal and not in a temporal sense. The relation between God and the work of his hands is essentially a causal relation and not a relation in space and time. The question, "Where was God before the creation?" is meaningless, since space is a "pure relation" (*idāfat mahda*) among created bodies.

Theological polemics, as we see, led to a conception of space as a network of relations a conception that shows a striking resemblance to Leibniz's idea about space. To be sure, it is al-

---

[79] Aristotle, *De caelo,* I, 9, 279 a.
[80] *Ibid.*
[81] Algazel, *Tahafot Al-Falasifat* (ed. by Maurice Bouyges, S. J.; Beyrouth, 1927), p. 53.

ways very difficult to assess any influence where thought processes are involved. Certainly, allowance has to be made for a
complete independence between similar conceptions, especially
if they are separated so much by time, space, and language. In
the case under discussion it is certainly wise to defer judgment
until a comparative study of the relevant texts establishes the
fact of an indubitable intellectual dependence or until historic-
biographic research proves the indebtedness without doubt. In
our case it is rather tempting to suspect such a dependence if it
is also noted that Leibniz's monadology shows a striking resemblance to the atomistic theory and occasionalism of the
Kalām, a famous Muslim school of thought, called also the "Mutakallimūn," or "Loquentes," as mentioned by Saint Thomas Aquinas. Details about their theory of space will be explained in
Chapter III. As far as our question of theological influence on
ideas about space is concerned, we have to stress the following
fact: It has been established that the theory of atoms in Islam
and the corresponding conception of space were originally of
a purely profane character and became adapted to an extreme
theistic dogma only during later stages of their development.
From a strictly historical point of view it must be admitted,
therefore, that the Kalāmic theory of space did not originate on
the background of religious speculations. However, it was this
background from which it drew, at the climax of its vitality, its
emotional strength of conviction. Our exposition of the Kalāmic
conceptions of space, in Chapter III, refers to this later stage,
the "orthodox" Kalām, which once was defined as "the science
of the foundations of the faith and the intellectual proofs in
support of the theological verities." [82]

[82] Sir Thomas Arnold and Alfred Guillaume, *The legacy of Islam* (Oxford
University Press, London, 1949), p. 265.

THE EMANCIPATION OF THE SPACE

CONCEPT FROM ARISTOTELIANISM

In Aristotle space is identified with place and defined as the adjacent boundary of the containing body. This definition, it is clear, is in line with Aristotle's fundamental assumption of the impossibility of a vacuum. Theophrastus' critique of his master's doctrine had no immediate import for the development of physical thought. In this respect the teachings of Strato of Lampsacus, the *"physikos,"* the successor of Theophrastus, exerted a greater influence. Strato's realistic attitude, probably a result of the Alexandrian climate of opinion, led him to divest the Aristotelian system of its transcendental elements and to seek for a compromise with everyday experience. He came to the conclusion that a vacuum is not an absolute impossibility, but may exist, and in fact does, in matter itself, forming minute interstices between the material particles. That this deviation from Aristotle's teaching induced Strato to revise the Aristotelian conception of space seems to be highly likely, if we remember that Strato wrote a book on the vacuum, which is now lost, but which was still known to Simplicius. Regrettably little, except some experimental features of this work, are referred to in later scientific literature, as for instance in the writings of Heron, who construed the penetration of light rays assumed to be of corporeal nature, and of heat through water as a proof of the existence of small vacua inside matter, as Strato had contended. In Heron's *Pneumatics* we find the assertion that continuous vacua can be produced, though only by artificial means, whereas in nature only small, discontinuous vacua can exist.

The first major contribution to the clarification of the concept of absolute space was made by Philoponus, or John the Grammarian, as he is often called (*fl. c.* A.D. 575). Philoponus is well known as the forerunner of the so-called "impetus theory" in mechanics, which was the subject of profound investigation during the fourteenth century and which became in its later development the main point of departure for Galilei's formulation of the basis of modern dynamics. We shall have occasion to see how Philoponus' revision of the Aristotelian conception of space is intrinsically connected with his impetus theory. He begins by pointing out an inner inconsistency in Aristotle's theory of space. To Aristotle, place is the adjacent boundary of the containing body, provided this containing body itself is not in motion. If, for example, we hold a stone in a current of water, the constantly changing envelope of water clearly is not the "place" of the stone; otherwise the motionless stone would change its place continuously, which is self-contradictory. The stone's place must therefore be the inner surface of the first immobile containing body, as, for instance, the river bed.

Philoponus now asks what actually is the place of the sublunary world of matter, subjected to generation and decay. According to Aristotle it is the concave surface of the first celestial sphere, the orbit of the moon. But, says Philoponus, this surface itself is constantly rotating and therefore not immobile; on the contrary, a certain part of this surface successively touches other parts of the contained matter, even if these parts themselves happen not to be moving. To ascribe the place of our changing world to one of the higher spheres is of no avail, since all of them are in rotational motion. Philoponus rejects the argument that rotation about a fixed axis or a fixed point is not local motion, since the sphere as a whole always occupies, so to say, the same place.[1] Philoponus concentrates upon a fixed part of the rotating sphere and shows how this part occupies different places in the course of time. Hence he concludes that Aristotle's definition of "place" leads to a cul-de-sac and must be rejected. The definition not only makes it impossible to determine the place of the sublunar world, but provides no answer to the question of the place or the space in which the outermost sphere is moving, for moving it certainly is.

Philoponus was not the first to notice this difficulty in explaining the motion of the last sphere consistently with Aristotle's principles. In fact, Aristotle's statement, "It is clear that there is neither place nor void nor time beyond the heaven," [2] began to be a subject of serious doubt. As adopted by most of the commentators, the usual solution to this question was to point out that the place of each individual part of the rotating sphere was determined by the parts of the same sphere that were contiguous to it. But, contends Philoponus, if this is the case, what parts of the rotating sphere actually "change their places" during the rotation?

A solution of this problem was offered, for instance, by Themistius, whose *Paraphrasis in libros quatuor Aristotelis de*

---

[1] In modern words, being a surface of constant curvature it is transferable into itself.

[2] Aristotle, *De caelo*, 279 a. 12.

caelo is extant in a Hebrew translation and also in a later Latin translation of this Hebrew text by Moses Alatino.[3] In his attempt to overcome the difficulty, Themistius lands himself in a vicious circle, for what he says boils down to the statement that the place of the outermost sphere is the convex surface of the sphere of Saturn, as much as the place of Saturn is the concave inner surface of the last sphere.[4]

Inconsistencies of this kind proved to Philoponus that a new definition of "place" or space was necessary. According to him, the nature of space is to be sought in the tridimensional incorporeal volume extended in length, width, and depth, different altogether from the material body that is immersed in it. "Space is not the limiting surface of the surrounding body . . . it is a certain interval, measurable in three dimensions, incorporeal in its very nature and different from the body contained in it; it is pure dimensionality void of all corporeality; indeed, as far as matter is concerned, space and the void are identical."[5]

However this identification of space and void does not assume the existence of a void as such *"in actu."* The void, although a logical necessity, is always coexistent with matter. Void and body are two inseparable correlates, each of them requiring the existence of the other. As soon as one body leaves a certain part of space, another body "replaces" the first. A certain region of space can receive different bodies in succession without taking part in the motion of the occupying matter. Philoponus' phoronomy is completely analogous, as Duhem[6] points out, to Aristotle's doctrine of substance and form, where one form is succeeded by another continuously, so that substance is never void of form. Just as matter successively receives one form after another, so a section of space may be occupied by one body after another, space itself remaining immobile.

    [3] Moses Alatino, *Themistii in libros Aristotelis De caelo paraphrasis, Hebraice et Latine* (ed. S. Landauer; Berlin, 1902).
    [4] Cf. Simplicius, *Physics*, p. 589.
    [5] H. Vitelli, ed., *Ioannis Philoponi in Aristotelis physicorum libros quinque posteriores commentaria* (Berlin, 1888), p. 567.
    [6] P. Duhem, *Le système du monde* (Paris, 1913), vol. 1, p. 381.

It is clear that this rather abstract notion of space is incompatible with Aristotle's dynamics, for Philoponus conceives space as pure dimensionality, lacking all qualitative differentiation. Space can no longer be conceived as the efficient cause of motion, compelling the body to move to its "natural place." "It is ridiculous to pretend that space, as such, possesses an inherent power. If every body tends to its natural place, it is not because it seeks to reach a certain surface; the reason is rather that it tends to the place which was assigned to it by the Demiurgus."[7]

Such concepts as "up" and "down," whose justification Philoponus does not deny, are no longer an intrinsic quality of space or place, but owe their validity to purely geometrical arrangement on the one hand, and to cosmological-theological predestination, as assigned by the Demiurgus, on the other. As to the first of these it should be pointed out that Philoponus, following Aristotle, accepts the fundamental tenet of the finiteness of the universe. Since matter is finite, its correlate, space, which is indissolubly connected with it, must be finite as well. Thus the universe possesses a final boundary, a last sphere, which determines its "upper" regions. The center, which as decreed by the Demiurgus is occupied by the earth, is by definition the direction of the "down." A body falls "down," not because its new place exerts such a strain as to direct it to its "natural place," where this strain ceases to operate, but rather because it possesses an inherent tendency to reach the place assigned to it by the Demiurgus. It is this tendency, inherent in the moving body and not in the medium or in space, that corresponds to the "impetus" in the case of forced motion. Philoponus' explanation of the fall of heavy bodies shows a remarkable resemblance to the explanation of gravity suggested by Copernicus: "Equidem existimo gravitatem non aliud esse, quam appententiam quandam naturalem partibus inditam a divina providentia opificis universorum, ut in unitatem integritatemque suam sese conferant in forman globi coëuntes."[8]

[7] Reference 5, p. 581.
[8] *De revolutionibus orbium coelestium,* liber I cap. IX.

Iamblichus' theory of space, because of its influence on physical thought in antiquity, should be mentioned here. The theory stands in complete contrast to that of Philoponus. As Simplicius relates in detail,[9] Iamblichus defines place as a material force which sustains the body and holds it together, which raises what has fallen and brings together what has departed, filling their volume and surrounding them from all sides. Duhem[10] thinks it probable that Iamblichus was influenced by the writings of Archytas.

A profound inquiry into the nature of space is to be found in Damascius' treatise *Peri arithmou kai topou kai chronou,* of which Simplicius gives us a detailed account in the *Corollarium de loco*[11] of his *Commentaria* to Aristotle's *Physics.* The primary term in Damascius' inquiry is not place or space, but position or location, which is for him an inseparable attribute of the object. In fact, this notion has two meanings, one denoting the relative location of the different parts of the object and the other denoting the position of the whole in the universe. Speaking in modern terms, we find here, probably for the first time, the notion of the three degrees of freedom of a complex body as a whole as opposed to its internal degrees of freedom. Space is as different from position as time is different from motion. Just as time for Aristotle is the numerical measure of motion, so for Damascius space or place is the numerical measure of position. If we look upon position as a certain quality of the object, space makes it possible to determine this quality quantitatively. But the essence, the nature of this quality, is not accessible to geometric formulation. In the same way as every part of the universe has a "natural" position which is best for that part, so the whole universe has a "natural disposition," attained when all its parts are in their corresponding "natural" positions.

[9] Simplicius, *Physics,* p. 639.
[10] *Le système du monde,* vol. 1, p. 333. For a detailed discussion on Iamblichus' theory of space, see also Eduard Zeller, *Die Philosophie der Griechen* (Leipzig, 1881) vol. 3, 2, p. 706.
[11] Book IV, cap. IV.

In contrast to the traditional conception of place in Greek thought, Damascius holds that position is inseparable from the object, even when the object is in motion. Place was usually supposed to be capable of receiving different bodies in succession; but position, like any other attribute, was not directly transferable from one object to another. Just as a body, when changing its color from white to black, does not leave whiteness behind as an independent existent, so the position of a body in motion, although constantly changing, never becomes the position of another body, but ceases to exist when the moving body acquires a new position.

Damascius' conception of space as the geometric measure of position led him to an important conclusion which again does not conform to the traditional Peripatetic doctrine. It is the famous question whether motion presupposes rest. Aristotle, or whoever wrote *De motu animalium,* lays it down that whenever a body is in motion, there must be something that is motionless.[12] Duhem thinks that the writer of *De motu animalium* did not himself refer explicitly to this axiom as an argument for the immobility of the earth. Still, the majority of commentators undoubtedly saw in it a proof of the necessity of the earth's immobility. For example, Themistius in his *Paraphrasis in libros Aristotelis de caelo* says: "Sed conversio, immo omnis motus, super manente ac quiescente aliquo omnino celebratur. In iis autem, quae De Animalium Motu a nobis dicta sunt, monstratum est id, quod manet ac quiescit, illius autem partem esse non posse, quod super ipso movetur."[13] As a matter of fact, the author of *De motu animalium* seems himself to have been aware of the cosmological implications of his biological conclusion, as when he writes: "And it is worthwhile to stop and consider this dictum; for the reflection which it involves applies not merely to animals, but also to the motion and progression of the uni-

[12] Cf. Aristotle, *Movement of animals,* 698 b 10, trans. by A. L. Peck (Loeb Classical Library; Harvard University Press, Cambridge, 1937).
[13] Reference 3, p. 97.

verse. For just as in the animal there must be something which is immovable if it is to have any motion, so *a fortiori* there must be something which is immovable outside the animal, supported on which that which is moved moves." [14]

The question whether motion presupposes the existence of something immobile or not divided Aristotle's followers during the centuries. This question was held to be a problem in dynamics and to be different from the purely kinematic phenomenon of apparent motion which pertains to sense perception. The latter question was dealt with adequately by Euclid, who stated that an object which appears to be at rest to an observer who regards himself as being moved, would appear to be retrogressing to the same observer if he regards himself as being at rest.[15] Averroes, who expounds Aristotle's theory of space and motion in detail, supports the view that a concrete motionless body is a necessary condition for the existence of motion. The problem assumed a vital importance at the end of the thirteenth century when its implications for theology became apparent. In the council of the doctors of the Sorbonne, which took place under the presidency of Etienne Tempier in 1277, Averroes' interpretation was declared to be heretical, since the recognition of an absolutely immobile body, immobile even for the Creator of the universe, was thought to be incompatible with the fundamental ideas of Christian theology. Their belief in God's omnipotence forced the theologians to the conclusion that God could move the whole universe — which of course was thought to be of finite extent — through space. We shall later have an opportunity to discuss the physical implications of this condemnation of Averroistic conceptions of motion and shall show how the attempt to reconcile it with Aristotelian physics led to a new

[14] Aristotle, *Movement of animals*, 698 b 10.
[15] "Si aliquibus latis pluribus inaequali celeritate simul transportetur in easdem partes et oculus, quae quidem oculo aequali celeritate feruntur, videbuntur stare, tardiora vero in contrarium ferri, celeriora vero in praecedentia. — Si aliquibus latis appareat aliquid, quod non feratur, videbitur illud non latum retrorsum ferri." I. L. Heiberg and H. Menge, ed., *Euclidis opera omnia* (Leipzig, 1883–1916), vol. 7, *Optica*, p. 110.

interest in the problem of space and motion. At present we are interested only in the fact that the theologians of Paris accepted the doctrine of Damascius as the only orthodox doctrine.

For according to Damascius motion presupposes no immobile body. It is only our perception of motion which necessitates reference to something that is supposedly not in motion; we need the assistance of a fixed object if we are to distinguish motion from rest through the change of certain geometric measures. The absence of a motionless body does not preclude the possibility of local motion; it merely prevents our recognition of motion as such. A motion of the whole universe is therefore not impossible. It is also possible, says Damascius, that the heavens would continue in their habitual diurnal revolutions even if neither east nor west nor south existed.[16] Although this appears to be the earliest affirmation of the merely relative value of geographical or astronomical directions in space, we must be careful not to exaggerate the point. For Damascius still holds to the traditional doctrine of natural places, which remain immobile and fixed, that is, independent of the real motion of the concrete parts of the universe.

If we were to translate Damascius' conception into modern terms, we might say that his set of natural places is identical with an extended field whose coördinates stand in a one-to-one correspondence with the material parts of the universe. This field is invariant and independent of the actual motion of the universe, but, as determining the final cause of natural motion, it may be thought of as a regulative force, making for an increasing degree of perfection of the universe. In this sense natural place is characterized by Damascius as the "telesiurgus," or the driving force to perfection.[17] As the final system of reference for the actual positions of all mobile bodies, "natural place" does duty in this sense for Aristotle's outermost immobile sphere, the final containing concave surface. However, as the things

---

[16] Simplicius, *Physics,* p. 634.
[17] *Ibid.,* p. 601.

of the universe have not reached their natural places, this system of reference remains an ideal abstraction, of no avail for
the physical determination of the position of actual bodies.

One question has still to be answered: Does Damascius endow natural place with efficient as well as final causes? In other
words, does a natural place, in terms of his theory of space,
exert a direct and directive force on the material body to which
it corresponds? Duhem answers this question in the affirmative,[18]
basing himself on Simplicius' remarks that Damascius admired
Iamblichus' doctrine and saw in him a predecessor of his own
opinion.

Turning from these commentators of the Stagirite to Muslim
exponents of Aristotelianism and Greek philosophy in general,
we note that once the Arabian world became familiar with the
system of thought contained in the writings of the Greeks and
Syrians, the authority of Aristotle became paramount. As far as
our subject is concerned, only a few major deviations from the
system of "al-failasūf" ("the philosopher" par excellence) are
encountered: the atomistic doctrine of space in the Kalām, and
the theories of Al-Razi and of Abu'l Barakāt.[19] The Kalām may
be compared with scholastic philosophy of medieval Europe
not only because of its dialectic method in theological speculation but also because of its object of supporting a dogma by
discursive thought. Abu' l'Hasan al-Ash'arī of Bagdad and Abū'l-
Mansūr al-Maturīdī of Samarquand, both of the tenth century,
are usually named as the principal founders of the orthodox
Kalām, although its origin most certainly dates back to the ninth
century.[20]

In order to bring into special prominence the divine creative
act, the Kalām attributes to matter (as well as to space, as we

---

[18] Duhem, Le système du monde, p. 350.

[19] Cf. S. Pines, "Etudes sur Awhad al-Zamān Abu'l Barakāt al-Baghdādī,"
Rev. études Juives 3, 5 (1938).

[20] According to Ibn-Khaldūn it was Al-Bāqilānī (d. 1013) of Basra, the
most remarkable disciple of Al-Ash'arī, who introduced the concept of
atomism into the Kalām. See G. Sarton, Introduction to the history of
science (Baltimore, 1931), vol. 1, p. 706.

shall see immediately) only a transient existence of extremely short duration and range, requiring thereby a constant divine creative interference for the maintenance of coherence and continuity in the universe. Everything contained in this universe was conceived as being composed of atoms and accidents, and not of substance and properties as Aristotle taught. This atomistic doctrine seems at first sight to stand in opposition to the theological tenet of the Kalām; but as soon as the principle of causality or the Democritian concept of "necessity" is abandoned and replaced by the adoption of a transcendental principle of divine interference, the contrast dwindles away. The refutation of a mutual interaction among the atoms is also in accordance with Aristotle's argumentations[21] with regard to a consequential atomism. Such a revised atomism proves to be a most suitable ground for the extreme theistic philosophy of the Kalām. The doctrine of atomism was regarded as the first fundamental proposition in the system, as we see in chapter 73 of Moses Maimonides' *Guide for the perplexed*.[22] This work, written with the purpose of reconciling Aristotle with Jewish theology, serves as an important source for our information on Muslim philosophy in general and the Kalām in particular, although it does not treat the latter in an impartial way.

The atoms of the Kalām are indivisible particles, equal to each other and devoid of all extension. Spatial magnitude can be attributed only to a combination of atoms forming a body. Although a definite position (*ḥayyiz*) belongs to each individual atom, it does not occupy space (*makān*). It is rather the set of these positions — one is almost tempted to say, the system of relations — that constitutes spatial extension. Thus, for instance, according to Mu'ammar (or Ma'mar), one of the older advocates of the Kalāmic theory, two atoms, if connected, related, or attached (*indamma*) with each other, constitute length; four

[21] Aristotle, *De generatione et corruptione*, I, 9, 326 a.
[22] For an analysis of this chapter, see D. B. MacDonald, "Continuous re-creation and atomic time," *Isis* 9, 342 (1927).

atoms constitute length and breadth, that is, a two-dimensional spatial extension; a three-dimensional body is composed of a pile (*tabaka*) of two-dimensional extensions and contains consequently at least eight atoms.[23] As Isaac Israeli's *Liber de elementis* clearly shows,[24] the problem of reconciling spatial extension of bodies with the supposedly unextended nature of atoms was a much discussed topic already in early Muslim-Jewish natural philosophy.

In the Kalām, these rather complicated and surprisingly abstract ideas were deemed necessary in order to meet Aristotle's objections[25] against atomism on the ground that a spatial continuum cannot be constituted by, or resolved into, indivisibles nor can two points be continuous or contiguous one with another.

In view of the surprising resemblance between the atomistic theory of the Kalām and the monadology of Leibniz, as well as between the corresponding conceptions of extension and space, we face the interesting problem whether this is a mere fortuitous coincidence. We know with certainty that Leibniz read Maimonides' exposition of the Kalām in Buxtorfius' Latin translation of the *Guide*. The copy used by Leibniz shows many marginal notes written by his own hand, demonstrating what great inspiration he drew from reading the book. Foucher de Careil, one of the editors of Leibniz's works, adduces additional information on this point in his book *Leibnitz, la philosophie juive et la Cabbale* (Paris, 1861). M. Guttmann draws our attention in this connection to the following phrase in Leibniz's *Epistolae ad P. des Bosses*: "Substantia nempe simplex etsi non habeat in se extensionem, habet tamen positionem, quae est fundamentum extensionis." [26]

An important feature of the atomistic doctrine of the Kalām

---

[23] S. Pines, *Beiträge zur Islamischen Atomenlehre* (Berlin, 1936), p. 5.
[24] Isaak b. Salomon Israeli, *Sefer Hayesodoth* (Frankfurt a. M., 1900), p. 43 of the Hebrew text.
[25] Aristotle, *Physics*, IV, 6; VI, 4, 6; 213 b, 234 b, 237 a.
[26] Moritz Guttmann, *Das philosophische System der Mutakallimūn* (Breslau, 1885), p. 20.

is its affirmation of the existence of empty space. Empty space is not only a necessary presupposition for the possibility of motion — combination and separation among atoms explain the processes of generation and corruption — but the disjunctive character of empty space is also claimed as a necessary prerequisite for the separateness and independence of the individual atom. Consequential thought led the Kalām to the conclusion that space, as well as matter (and time), is of atomistic structure. Otherwise, that is, on the assumption of a spatial and temporal continuity, matter could be proved to be divisible ad infinitum, contrary to the first fundamental proposition. Discontinuity of space and time leads to the peculiar, but logical, explanation of motion as a series or sequence of momentary leaps: the atom occupies in succession different individual space-elements. Physical motion becomes thus a discontinuous process.

To be exact, the discrete structure of space, according to the theory of Kalām, can be inferred from the two premises (1) of the discreteness of time (the third fundamental proposition of the Kalām, according to the enumeration of Maimonides); (2) of the Aristotelian inference from the continuity of space to that of motion, and from the continuity of motion to that of time.[27] Since the consequent, according to the first premise, is denied, the formal application of the *modus tollens* leads to the conclusion that space is not continuous.

The atomistic theory of space gave rise to many complications. First of all, it became evident that differences in speed can no longer be attributed to the fact that the body which has moved through a larger distance had a greater velocity, but must be due to the circumstance that the "faster" body has been interrupted by fewer moments of rest. Fundamentally, only one common speed (frequency) lies at the basis of all physical processes. Had motion pictures or electric advertising with its stroboscopic illusions been an invention of the Middle Ages, the proponents of the Kalām would have faced no difficulties in

[27] Aristotle, *Physics*, IV, 1, 3.

finding suitable illustrations for their teachings.

That this peculiar conception of motion led to considerable complications was soon realized. Thus Maimonides argued:

"Have you observed a complete revolution of a millstone? Each point in the extreme circumference of the stone describes a large circle in the same time in which a point nearer the center describes a small circle; the velocity of the outer circle is therefore greater than that of the inner circle. You cannot say that the motion of the latter was interrupted by more moments of rest; for the whole moving body, i.e., the millstone, is one coherent body." They reply, "During the circular motion, the parts of the millstone separate from each other, and the moments of rest interrupting the motion of the portions nearer the center are more than those which interrupt the motion of the outer portions." We (Maimonides) ask again, "How is it that the millstone, which we perceive as one body, and which cannot be easily broken, even with a hammer, resolves into its atoms when it moves, and becomes again one coherent body, returning to its previous state as soon as it comes to rest, while no one is able to notice the breaking up of the stone?" Again their reply is based on the twelfth proposition, which is to the effect that the perception of the senses cannot be trusted, and thus only the evidence of the intellect is admissible.[28]

The argument of the revolving millstone, which was advanced by Maimonides with the obvious purpose of showing the inner inconsistency of the Kalāmic space theory, is but another version of the well-known problem which from the late fifteenth century on became renowned under the ambitious name of "rota Aristotelis." It was known, however, throughout the Middle Ages and led to many investigations into the structure of space and in some cases to a rejection of the traditional Aristotelian doctrine of continuity. The problem essentially is this: two concentric circles with different radii, rigidly connected to each other, move in such a way that each of them, during one complete rotation, rolls along a straight line (Fig. 1); how can these two lines be equal in length, being produced by circumferences of different radii?

[28] Moses Maimonides, *The guide for the perplexed*, trans. by M. Friedlaender (Pardes Publishing House, New York, 1946), chap. 73, p. 122.

As late as in the seventeenth century "interposed vacua" or infinitesimal "moments of rest" were postulated in order to solve the problem. It will be recalled that Galilei also discusses the problem in his *Discorsi e dimostrazioni matematiche, intorno a due nuove scienze*,[29] and that his treatment of the "infinite and indivisible" is reminiscent of the ancient teachings of the Kalām.

By assuming the discontinuity of motion the Kalām protected itself against Aristotle's famous attacks[30] on atomism, later repeated in another form in Al-Ghazālī's *Makāṣid-al-falāsifa*, ac-

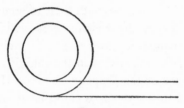

Fig. 1

cording to which the conception of a continuous motion in an atomic universe leads necessarily to a division of the indivisible and is therefore incompatible with an atomic theory of space and time.

Another important result of the theory of the Kalām was the rejection of a possible incommensurability between spatial data and a denial of the existence of irrational magnitudes (lines, etc.). If every spatial extension, say a line, is atomistic in structure, that is, composed of an integral number of atoms, clearly no incommensurable lines can exist and no irrational measures can be conceived. The Pythagorean discovery of the irrational was in the view of the Kalām but an unhappy chimera.

Very little is known about a possible influence of the Kalāmic space theory on scholastic thought in medieval Europe. As a

[29] Edizione Nazionale, p. 70 and 96; Dialogues concerning *two new sciences* trans. by H. Crew and A. de Salvio (Dover, New York, 1952), p. 22, 51.
[30] Aristotle, *Physics*, VI, 232 a, 233 b.

pure speculation, which ignores any scientific relevance of sense data and consequently is extremely averse to experiment and observation, it remained necessarily restricted in its physical contents to the discussion of motion. It is possible, therefore, to trace its influence only in scholastic investigations into the nature of space, motion and the continuum, a subject which was one of the major attractions for fourteenth-century schoolmen. But even here we have little evidence of any direct dependence and all our conclusions are only conjectural.

To be sure, it is an established fact that the works of Al-Ghazālī[31] and of Maimonides, with their references to the atomistic space theories of the Kalām, were widely read by the scholastics. The *Guide for the perplexed* was most probably translated into Latin as early as in the middle of the thirteenth century at the court of the Emperor Frederic II.[32] Is it possible that the atomistic theory of space could have escaped the attention of William of Auvergre, Francis of Sales, Vincenz of Beauvais, Albertus Magnus, Thomas Aquinas, Duns Scotus, and many others who were well acquainted with Maimonides' *Guide*?

On the other hand, the problem whether spatial magnitudes — lines, areas, and volumes — were constituted of indivisibles ("compositio ex indivisibilibus") or of points ("compositio ex punctis") was widely discussed in the course of the fourteenth century. The majority of the schoolmen retained the Aristotelian doctrine according to which the continuum is characterized as being composed of parts which themselves can be divided and subdivided ad infinitum ("continuum est constitutum ex quantibus divisibilibus in alias quantitates"); it is not composed of indivisibles. Aristotle recognized that this regression of an endless division introduces the concept of infinity. But this conception — in contrast to the notion of an infinitely extended space — requires, in his view, only the idea of a potential infinite.

[31] Al-Ghazālī's works were translated into Latin in the twelfth century by Dominic Gundisalvi.
[32] See Graetz, *Monatschrift* (Jan.-June 1875).

Against those, like Nicolaus of Autrecourt or Henry of Harclay, who contended that spatial extension is composed of dimensionless and indivisible points, the following principle was quoted: "Indivisibile indivisibili additum non facit maius," a principle that expressively contradicts the Kalām conception of space. Nicolaus of Autrecourt's reasoning, incidentally, rests on his objection to the Aristotelian conception of space as a plenum. If space were a plenum, he contends, one of three "inconveniences" would necessarily follow: either rectilinear motion would be impossible, or two bodies would be at the same place at the same time, or the motion of one body would imply the motion of all the other bodies in the universe.[33] Reason and experience demand, therefore, in his view, the postulation of the existence of vacuities and the rejection of the Peripatetic theory of space.

The greatest resemblance to the Kalām theory of space is exhibited by the teachings of Duns Scotus' disciple Nicolaus Boneti,[34] who advanced an extreme atomistic theory of spatial extension. Unfortunately, very little has been published about his teachings.

After this digression to questions relating to the continuity of space, let us resume the main line of our story and explain how an intrinsic critique of the traditional Peripatetic conception of space led gradually to far-reaching consequences, culminating finally in the emancipation of the concept of space from the doctrine of substance and accident.

So far our discussion has been confined mostly to the major theories of space in Antiquity and their recurrences in scholastic thought. They may be classified under three headings: the atomistic view (with its emphasis on the physical character of space), the Platonic view (with its emphasis on mathematics), and finally the Aristotelian view (with its ontology). With regard to our special problem, as well as generally, the early periods of

[33] Julius Rudolph Weinberg, *Nicolaus of Autrecourt* (Princeton University Press, Princeton, 1948).

[34] See Anneliese Maier, *Die Vorläufer Galileis im 14. Jahrhundert* (Storia e Letteratura, Rome, 1949), p. 177.

medieval thought exhibit a strong inclination toward Platonism which gives way in the late scholastics to Aristotelianism, until with the dawn of modern science war is declared on Peripatetic thought and Neoplatonism becomes the main ingredient in Italian natural philosophy. The early Middle Ages contribute little to the development of the concept of physical space. On the other hand, and far more important, is the physical thought of the later period in which Aristotle is of supreme influence.

It will be recalled that place is the concave surface of the containing body and is by its nature immobile. In the light of this definition let us refer to the following passage in Aristotle's *Physics:* "Of things which are in motion some are moved by the actualizing of their own inherent potentialities, and others only by being involved in the movement of something else in which they inhere." [35] For example, a nail in the side of a ship does not move by itself (*kath' auto*), but is moved *per accidens* (*symbebēkos*), without changing its place. Here we come upon the first conceptual difficulty in Aristotle's doctrine of space, a difficulty that became one of the major problems of medieval physics. It is this: If space is the concave surface of the containing body, and motion is change of space, how can the concept of "motion *per accidens*" be reconciled with these definitions? Looking at the problem in modern terms, it is clear that Aristotle was fully aware that motion can be inferred only with reference to a second body, that is, by the choice of an immediately surrounding body as a reference system. Thus Aristotle raised a difficulty that has baffled many thinkers throughout the ages. Sextus Empiricus in his *Against the physicists* was already struggling with this obvious contradiction. He writes:

These motions, then, are omitted from their description; but there is also another more surprising kind of transitional motion, in which the moving object is conceived as not going out from the place wherein it is either as a whole or part by part; and this too is omitted from their definition, as is obvious at once. And the peculiar character of this motion will be more evident when we have explained it by an

[35] Aristotle, *Physics,* IV, 4, 211 a 18.

example. For if we should suppose that, when a ship is running before the wind, a man is carrying an upright rod from the prow to the stern and moving at the same speed as the ship, so that in the time in which the latter completes the distance of a cubit in a forward direction, in an equal time the man who is moving in the ship passes over the distance of a cubit in a backward direction, then, in the case thus supposed there will certainly be transitional motion, but the moving object will not go out from the place wherein it is either wholly or in part; for the man who is moving in the ship remains in the same perpendicular both of air and of water owing to the fact that he is borne just as far forward as he seems to proceed backward. It is, then, possible for a thing which does not quit the place wherein it is wholly or in part to move transitionally.[36]

In these words Sextus Empiricus attacks Aristotle's definitions of place and motion, but he fails to see the possibility of adopting as a system of reference some distant body to which the body under consideration is brought into spatial relation. This would not only have eliminated the difficulty with which he is concerned, but would also have it made possible to correlate the motions of various bodies under one common aspect. But such a step would have obliged him to reject Aristotle's definition of place as the surface of the *adjacent* body. The authority of Aristotle was too great for such a radical change. Even William of Occam, the revolutionary nominalist of the fourteenth century, considered it necessary to adhere to Aristotle's definition of space:

Various explanations are suggested by various people in order to maintain the immobility of place. Thus some say that place has two aspects, namely, that which is material in place, viz., the surface of the containing body; secondly, that which is formal in place, viz., its order with regard to the universe (*ordo ad universum*). This order in relation to the universe, however, is always immobile. For place, with regard to its formal aspect, cannot be moved either for itself or *per accidens* . . .[37]

By way of illustration Occam refers to the classical example of the ship lying at anchor.

[36] Sextus Empiricus, *Against the physicists*, II, 55.
[37] William of Occam, *Summulae in libros physicorum* (Bologna, 1494).

Although new masses of water are constantly coming up around the ship and although the ship does not always occupy the same order in relation to the parts of the river, as these are constantly moving, still, in regard to the river as a whole, the ship rests in the same place as long as it is at anchor . . . If you are at rest, and even if all the air around you, or any body which surrounds you, is moving, you are always at the same place; for you are always at the same distance from the center and the poles of the universe. With regard to these the place is therefore called immobile.[38]

We appear to have here for the first time the introduction of distance for the identification of place. Thus immobility of a given place was reduced to the constancy of distance from a given reference body, or from a set of such bodies. As Occam's words "the same distance from the center and the poles of the universe" indicate, this reference body was usually the outermost sphere of Aristotelian-Ptolemaic cosmology.

So we come to the second problem that arises within the framework or Aristotelian physics. The outermost sphere in the Aristotelian universe was regarded as moving with constant angular velocity, but was itself without place,[39] being uncontained in any further sphere. We have already seen that this difficulty was the occasion for much subtle discussion.[40]

All attempts to reconcile the obvious contradiction between the notion of absence of place (or space) for the last sphere[41] and the assumption that it moves (and therefore changes its place, according to Aristotle's definition of motion) were doomed to failure. It remained a major problem in scholastic philosophy until Copernicus finally came to the conclusion that the two ideas were irreconcilable, and that at least one of them would have to be rejected. Either the definition of "place" had to be revised, or the dogma of the motion of the outermost celestial

[38] *Ibid.*

[39] Aristotle, *Physics*, IV, 5, 212 b 10, for instance.

[40] See p. 37 and also p. 53.

[41] Dante, *Paradiso*, XXVII, 109, with poetic license, suggests a theological solution of the problem:

> E questo cielo non ha altro dove
> Che la mente divina.

sphere had to be repudiated. As we know, Copernicus preferred the second alternative. That this problem was really one of the major incitements to Copernicus' drastic revision of the accepted cosmological conception may be seen from various remarks in his *De revolutionibus orbium caelestium* (1543). In chapter V of the first book he says: "Cumque caelum sit, quod continet et caelat omnia, communis universorum locus, non statim apparet, cur non magis contento quam continentj, locato quam locantj motus attribuatur." [42] Copernicus contends that it would be much simpler to attribute motion to the contained body than to the container, for this evidently would solve the problem. In chapter VIII of the same book[43] he goes so far as to call it "absurd" to attribute motion to the containing last body. He writes: "Addo etiam, quod satis absurdum videretur, continenti sive locanti motum adscribi, et non potius contento et locato, quod est terra." And when, in chapter X of the first book, Copernicus gives a preview of his new cosmology, he feels justified in saying: "Prima et suprema omnium est stellarum fixarum sphaera, se ipsam et omnia continens, ideoque immobilis; nempe universi locus, ad quem motus et positio caeterorum omnium syderum conferatur." [44] It seems to us clear that the word *ideoque* (therefore) indicates that to contain both itself and all other bodies implies the absence of motion. It is not usually noted that the Copernican revolution was the outcome, in part, of the solution of a difficulty involved in the Aristotelian definition of place or space. We are well aware that the problem of the "placeless" motion of the outermost sphere was not the only factor that led Copernicus to his new conceptions. Copernicus' way of solving this problem had been suggested already by Alexander Aphrodisiensis, who according to Narboni's *Kawwanot ha-Pilosophim* conceived of an immobile outermost sphere that does not exist in place; also, as we have seen on page 37, additional spheres have been proposed to con-

[42] F. and C. Zeller, ed., *Nicolai Copernici Thorunensis De revolutionibus orbium caelestium libri sex* (Oldenbourg, Munich, 1949), p. 14.
[43] *Ibid.*, p. 20.
[44] *Ibid.*, p. 25.

tain the sphere of the fixed stars; but in all these cases we do not know of any major change in the cosmological conceptions involved.

Concerning the first of the two alternatives mentioned above, namely, the rejection of Aristotle's definition of place, it was adopted more than 150 years before Copernicus. It was a step that led to drastic revisions of the whole of Aristotelian physics. As much a revolution as that of Copernicus, it failed to bear fruit owing to adverse conditions of a political and religious nature. We are speaking of Hasdai Crescas' critique of Aristotelian physics in his *Or Adonai* (*c.* 1400). If we are to see Crescas' contribution in true historical perspective, we must take up again the question of space outside the universe.

Aristotle's doctrine provided a clear and accurate definition of place, while the rival doctrine, as expounded by the atomists, and later, leaving out of account ancient Pythagorean lore, by the Stoics and Philoponus, omitted to give a strict definition of space or place, taking space as a more or less primitive concept in the construction of the system. As a matter of fact, the intuitive conception of a vastly extended space, surrounding the material universe, seems to have been dormant all through the ages and may be found underlying even the most conservative doctrines of theological cosmology in the Middle Ages. So, for example, among the errors condemned in 1277 there is this: "Quod Deus non possit movere Coelum motu rectu. Et ratio est quia tunc relinqueret vacuum." [45] In terms of Aristotelian physics the very idea of motion of the universe as a whole is absurd and senseless. For motion presupposes place — a place in which the moving object is and a place to which it tends. But in Aristotelian physics place by definition is found only inside the universe.

In order to illustrate the way in which fourteenth-century thought struggled with the problem of outer space, we quote in

[45] Denifle-Chatelain, *Chartularium Universitatis Parisiensis* (1889–97), vol. 1, p. 546.

detail a passage from Richard of Middleton's *Super quattuor libros sententiarum quaestiones subtilissimae.*

I answer that God could move the outermost sphere (whether by creating or not creating space outside it) in rectilinear motion, although it would be impossible for any power whatever to apply such motion to a body taken in itself and as a whole as far as there is no space outside it. From this (it may be inferred) that if only a single angel existed, God could not move him in such a motion, if it were not by creating space outside him or around him; yet God could move any body in rectilinear motion, even if there were no space outside it, on condition that the motion is partial and accidental. Likewise, if there were a hole in the empyrean sphere and if the littlest man had a lance the lower end of which he drove with a rectilinear motion towards the outermost surface of the empyrean heaven, he would cause a certain part of the lance, in the course of its motion, to pass through the last sphere, although outside this sphere no space exists. So I say that God, if He moved in proper rectilinear motion a part of the empyrean heaven towards the earth, while the dimension and quantity of the former remain unchanged, would cause another part of heaven to move in a rectilinear motion, albeit not in space. Thus too it is clear, that He can move the whole of heaven in rectilinear motion by the rectilinear motion of that part which He causes to move in space.[46]

[46] Richard of Middleton, *Super quattuor libros sententiarum quaestiones subtilissimae,* p. 186. The Latin text is also quoted in Alexander Koyré, "Le vide et l'espace infini au XIVme siècle," *Archives d'histoire doctrinale et littéraire du Moyèn Age* (1949), p. 71: Respondeo quod Deus posset movere ultimum coelum (sive creando spacium extra ipsum sive non creando) motu recto, quamvis enim eandem rem per se, et secundum se totam impossibile sit moveri motu locali recto, per quamquamque potentiam nisi extra ipsam sit aliquod spacium (unde si nulla creatura esset nisi unus angelus, Deus non posset ipsum angelum tali motu movere nisi in quantum posset creare aliquod spacium extra ipsum, vel circa), tamen per accidens vel secundum partem Deus posset movere corpus aliquod motu locali recto, quamvis extra ipsum nullum esset spacium, inde si esset aliquod foramen in coelo empyreo, et minimus homo habeat lanceam, impellendo per motum rectum, partem lanceae inferiorem versus ultimam superficiem coeli empyrei, faceret, quod lancea motu recto transcenderet quantum ad aliquam sui partem, ultimam superficiem coeli empyrei, quamvis extra ipsam nullum sit spacium, sic diso, quod Deus si moveret motu proprio recto unam partem coeli empyrei usque ad terram, figura coeli et quantitate salvis manentibus, faceret quod alia pars coeli moveretur motu recto; quamvis non in aliquo spacio. Sic ergo patet, quod posset totum coelum movere motu recto per rectum motum illius partis quam moveret in spacio.

Aristotelian physics in general and his theory of space in particular became subjected to penetrating and detailed criticism in the fourteenth century. Henri de Gand, Richard of Middleton, Walter Burleigh, and Thomas Bradwardine discussed the controversial problem of space and void extensively. At the same time, however, it must not be forgotten that revisions and criticisms were advanced mostly from the theological point of view. Nor were the revisions incorporated into a consistent system; often they were adopted merely as probable assumptions as for example the theory of the vacuum of Nicolaus of Autrecourt.[47]

In Hasdai Crescas we encounter an independent thinker who deals with his opponents only after having presented their case as objectively as possible. His critique of Aristotelian physics, though reared on the foundations of Jewish orthodox theology, does not confine itself to mere assertions or refutations, as was the case with the Kalām in Muslim philosophy. Crescas succeeds in clearly pointing out inherent contradictions and inconsistencies and only then proceeds to a new formulation and a revised conception. In proposition I, part II of his *Or Adonai*[48] Crescas refutes Aristotle's definition of place as the adjacent surface of the containing body by pointing out the many absurdities to which it necessarily leads. First of all, Aristotle's definition cannot be consistently applied to the heavens. In the *Physics*, where Aristotle discusses the problem whether the outermost sphere possesses a place or not, he says: "But heaven, as has been said, is not anywhere as a whole nor in a certain place, since there is no body embracing it; but as far as it is moved, it constitutes places for its own parts, since one part embraces another."[49] If "heaven" in this connection is understood, as Themistius understood it, as the outermost sphere, the meaning of the term "place" when applied to this outermost sphere is different from its mean-

---

[47] Cf. J. R. O'Donnell, "The Philosophy of Nicholas of Autrecourt," *Medieval Studies 4*, 97 (1942).
[48] H. A. Wolfson, *Crescas' critique of Aristotle* (Cambridge, 1929), p. 199.
[49] Aristotle, *Physics*, IV, 5, 212 b 8–13.

ing when applied to the other spheres. If, on the other hand, we accept the interpretations of Avempace and Averroes, according to which the place of the celestial spheres is their center around which they rotate, we are landed in still another inconsistency. The celestial bodies would be adapted to abide in a place that is beneath them, for every body is naturally adapted to abide in its place; yet fire is not adapted to abide in a place beneath it. Again, a continuously extended body, for example, the atmosphere, raises a further difficulty. The proper place of the air as a whole is the concave inner boundary of fire. What then is the proper place of a part of air, which is surrounded by other parts of air? Is its place identical with the place of the air as a whole?

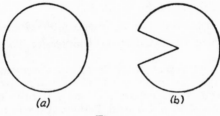

(a)   (b)

Fig. 2.

In that case, what happens to Aristotle's requirement[50] that place be equal to the object occupying it? On the other hand, if its place is the other parts of the surrounding air, the place then would not be distinct from what occupies it, so that here too Aristotle's requirements[51] would go by the board. Furthermore, the place of the part would not be a part of the place of the whole. In other words, Crescas' analysis shows that even for sublunar elements the requirements of "separateness," "surrounding," and "equality," all of which Aristotle holds essential for the concept of place, are incompatible. Further yet, the acceptance of Aristotle's definition of place leads to a paradox which is hinted at by Crescas and expounded in detail by his

[50] *Ibid.*, 211 a 27.
[51] *Ibid.*, 211 a.

pupil Joseph Albo.[52] It is this: the place of a part may be greater
than the place of the whole. Let us consider a spherical body, as
illustrated in Figure 2(a), and let us make a deep cleft in it,
as in Figure 2(b). It is evident that the new body, which obviously is only a part of the sphere, still has a greater "place" than
the whole sphere, a conclusion that contradicts common sense
and Euclidean geometry.

According to Crescas, any definition of place (or space) must
meet the requirement that the place of a body taken as a whole
be equal to the sum of the places of the parts into which it can
be broken up. Yet on the basis of Aristotle's definition, which
identifies place with the adjacent boundary, the "place" of a
cube is clearly smaller than the sum of the places of the little
blocks into which it can be divided.

Expounding his master's ideas, Joseph Albo criticizes Aristotle's
definition of "place" and writes in his Sefer-Ha-'Ikkarim:

> Furthermore it follows according to him that the place of the part
> is greater than the place of the whole. For if you remove part of the
> inside of a sphere, it will require a greater surface to bound it out-
> side and inside than when it is solid. Besides, it would follow accord-
> ing to him that one and the same body will have many places differing
> in magnitude. For if you divide a body into parts, each of the latter
> will require a greater place than before the division and the same is
> true if you divide the parts into other parts, and those again into parts.
> But this is contrary to the statement of Euclid in his book Concern-
> ing the heavy and the light,[53] where he says that equal bodies occupy
> equal places. But according to the Aristotelian hypothesis this is not
> true. For of two equal bodies the one that is divided will require a
> greater place than the other. All these difficulties follow from the
> opinion that place is an external bounding surface." [54]

Such are the main arguments that Crescas advances against
Aristotle's definition of place. They lead him to elaborate his

---

[52] Joseph Albo, Sefer Ha-'Ikkarim (Book of Principles) (ed. by Isaac
Husik, Philadelphia, 1929), vol. 2, p. 105. Cf. Dogmas II, 17 (Sonzino,
1485).

[53] About this spurious work see Steinschneider's article "Euclid bei den
Arabern," Zeitschrift für Mathematik, hist. litt. Abt., Bd. 31 (1886).

[54] Albo, Sefer Ha-'Ikkarim, p. 106.

own view, according to which space is a great continuum of infinite dimensions, an immovable void, ready to receive matter. "The true place of a thing is the interval between the limits of that which surrounds." [55] He adopts Aristotle's tentative, but finally rejected, definition that place is "some kind of dimensional extension lying between the points of the containing surface." [56] The grounds of Aristotle's rejection of this view is that it contradicts the requirements of separateness and immobility of place. By identifying this "dimensional extension" with the vacuum which becomes place when it contains a body, Crescas proceeds to show that Aristotle's arguments do not hold. His answer to Aristotle's argument that since all bodies move, if space were the interval of a body, space would be moving in space, is that there are no various spaces; space is one, infinite and immovable. By admixture of matter the infinite void becomes the extension of physical bodies.

Crescas' definition of space not only places him in opposition to Aristotelian physics, but makes him the first proponent of the reality of the vacuum in Jewish philosophy. Just as atomism, even in its Muslim theological form, was never endorsed by Jewish philosophers, unless exception is made of Abraham ibn Ezra, so the possibility of a void was always discarded by Jewish philosophical thought until Crescas became its great advocate. One reason for this traditional Jewish attitude toward the problem of the void was Aristotle's immense influence on Jewish thought in the Middle Ages; another was the empirical point of view adopted by Jewish thinkers.

Crescas challenged this attitude by refuting one after another of Aristotle's arguments against the existence of a void. In his refutation of Aristotle's argument that the existence of a void would preclude any motion, Crescas intimates his view on the physical structure of space; and hence his refutation is of special interest for us. According to Aristotle, the medium is an indispensable

[55] Crescas, *Or Adonai*, proposition I, part II; see reference 48, p. 195.
[56] Aristotle, *Physics*, IV, 4, 211 b 8.

condition for motion in that it serves constantly as both the *terminus a quo* and the *terminus ad quem* for the moving body whose natural motion is toward its natural place. Crescas tries to show that motion is not dependent on the existence of a medium. The first step in his argument is the assertion that weight and lightness are intrinsic qualities of the bodies and independent of any medium. "All moving bodies have a certain amount of weight, differing only *secundum minus et majus*," [57] upward motion being the result of pressure exerted upon bodies by others that are heavier.

In dispensing with the notion of an inner tendency in the elements to reach their natural places, Crescas revives certain views, particularly those of the atomists, according to which differences of weight are the result of differences in the internal structure of bodies. But these ideas are brought forward by Crescas merely as hypotheses. He does not follow them through to their logical conclusion and rejection of natural places altogether. He mentions them merely to show that the medium may be dispensed with as a cause of motion. To him, even if a more conservative view be taken and the motion of matter be regarded as an inner striving of the elements toward their natural places, the medium is under no circumstances the efficient cause of motion. For the parts of a vacuum can exhibit no differentiation in their constituent nature; for the vacuum is one homogeneous continuum. Yet, with respect to relative distances of these parts from the lunar sphere (the periphery) and the earth (the center), they do show a differentiation, but one merely of external relation. In this way Crescas comes astonishingly near to the idea of *actio in distans*. "Thus when fire moves from one part of the vacuum into another in upward motion, it is not because it tries to escape one part of a vacuum in order to be in another, but rather in its endeavor to get nearer to its proper place, which is the concavity of the lunar sphere, it naturally has to leave those remote

[57] Reference 48, p. 185.

parts of the vacuum and occupy the parts which are nearer to its proper place." [58]

The void or space, which according to Crescas becomes material extension when occupied by matter, is conceived by him as infinite in extension. Thus Crescas inaugurates a new front in his struggle with Peripatetic physics, according to which the universe is finite and limited. Aristotle's demonstration of the impossibility of infinity is set forth in *De caelo*,[59] in a classical passage which was plausible to the point of hypnotizing medieval thought.

Upward and downward motions are contraries, and contrary motions are motions to opposite places; and if one of a pair of opposites is determinate, the other must also be determinate. But the center is determined, for wherever the downward-moving body may come from, it cannot pass farther than the center. The center then being determined, the upper place must also be determined; and if their places are determined and limited, the bodies themselves must be limited.

Aristotle attempts also in the *Physics* and in the *Metaphysics* to prove the impossibility of both a corporeal and an incorporeal infinite extension. An account of Crescas' refutation of all these arguments would lead us too far afield, and we cannot do better than refer the interested reader to Wolfson's book.

As in his previous arguments, here too in his refutation of Aristotle's view of infinity, Crescas undermines his opponent by sound arguments and does not confine himself merely to expressing his disagreement. Accordingly, as the first proponent of infinite homogeneous space, Crescas made an outstanding contribution to the history of scientific thought. For he not only turned his back on Aristotelian conceptions, but through strict logical reasoning anticipated some fundamental ideas of sixteenth- and seventeenth-century physics. It was a great misfortune that he was never able to bring his ideas to their full flower. Political instability in Spain in the fifteenth century put an end to the intellectual activities of Catalonian Jews.

[58] *Ibid.*, p. 402.
[59] Aristotle, *De caelo*, I, 273 a 10.

Crescas' theory of space solved the problem of the outermost sphere: The infinite vacuum provides this sphere with space, so that its eternal rotation becomes a special kind of local motion and the sphere ceases to be the final limit and boundary of space.

Crescas' solution of the problem was not the only one advanced in the beginning of the fifteenth century. Nicholas of Cusa offered another. In his view, universal motion has no center, since in terms of his principle of the *"coincidentia oppositorum"* the absolute minimum must coincide with the absolute maximum. But God alone may be thought of as the absolute maximum of existenee, so that Cusanus comes to the conclusion: "Qui igitur est Centrum mundus? scilicet est Deus benedictus, ille est Centrum terrae, et omnium sphaerarum." [60] However, from the purely physical point of view, the identification of the center of the universe with its circumference is an obvious absurdity. To Cusanus the world has neither a center nor a circumference. "Quia minimum cum maximo coincidere necesse est. Centrum igitur mundi coincideret eum circumferentia. Non habet igitur mundus circumferentiam." [61] So it is clear that the earth is not the center of the universe or of space. "Terra non est centrum mundi." [62] The manner in which Cusanus goes on to derive the motion of the earth, thereby anticipating certain ideas of the Copernican theory, is not part of our subject. But it is important for us to note that the absence of a body absolutely at rest (the earth) does away with the possibility of absolute motion and absolute space. It is this relative character of position and motion that brands Cusanus' theory of space as modern. Another modern feature is its rejection of the idea that a hierarchy of values rules different regions of space. Of Aristotelian origin, the idea is im-

[60] Nicholas of Cusa, *De docta ignorantia*, II, 11; see A. Petzelt, ed., *Nicolaus von Cues, Texte seiner philosophischen Schriften, nach der Ausgabe von Paris 1514, sowie nach der Drucklegung von Basel 1565* (Kohlhammer, Stuttgart, 1949), vol. 1.
[61] *Ibid.*, 21.
[62] *Ibid.*

plied in the doctrine of physico-moral parallelism. As is well known, Aristotelian biology assigns to the upper parts of the human body a greater degree of nobility than to its lower parts. In consequence of this conception, as well as that of the parallelism between macrocosm and microcosm, the terms "high" and "low," though primarily purely geometric notions of spatial orientation, came in most languages to stand for distinctions of value.[63] The conception of a spatial hierarchy of values found its most perfect expression in Dante's *Divine comedy*, which from this point of view is a spatial metaphor of the gradations of sin and blessedness. How far this anthropomorphic conception became an integral part of medieval natural philosophy can be illustrated by the fact that Nicolaus of Autrecourt had to renounce his untimely thesis: "Quod non potest evidenter ostendi nobilitas unius rei super aliam." [64]

Cusanus, objecting the spatial hierarchy of values, states explicitly: "Neque dici debet, quod quia terra est minor sole et ab eo recipit influentiam, quod propterea sit vilior." [65] To Cusanus the earth is certainly not the smallest celestial body, the moon and Mercury being smaller; nor can any conclusion be drawn from the fact that the earth depends on the sun since the earth as a celestial body influences also in some degree the sun and its region.

The rejection of a spatial hierarchy of values is the logical conclusion of a more general principle which Cusanus advances in his *Docta ignorantia:* wherever in the heavens anyone may be placed, it would seem to him as if he were the center of the universe. This statement is evidently a rudimentary expression of the so-called "cosmological principle" of modern science as

[63] The designations "right" and "left" ("dextra," "sinistra") have their origin in a somewhat opposite development: the "propitious" or "faithful" (Hebrew: "yamīn") hand became the "right," the "sinister," malignant, the "left." A reference to the widespread belief that the left side is ill-omened is encountered in the Ebers Papyrus, the famous document on early Egyptian medicine, dating most probably from 3400 B.C.
[64] Denifle-Chatelain, *Chartularium Universitatis Parisiensis*, II, 544.
[65] *De docta ignorantia;* see reference 60, p. 106.

far as the spherical symmetry of space is concerned. The general validity of the principle that the universe presents the same aspect from every point (and according to a modern school of cosmologists also at every time), except for local irregularities, is accepted in modern science as a necessary condition for the repeatability of experiments, since space and time are the only parameters which, at least in principle, are beyond the control of the experimenter and cannot be reproduced at his will. Since this postulate in modern cosmology — not only with respect to the purely geometric aspect of space, but also with regard to its kinematic and dynamic aspects — has gained so much importance recently, it is not without interest to note that in Cusanus' writings we encounter, probably for the first time, an explicit enunciation of its spatial implications. If there were any justification for regarding Nicholas of Cusa as marking the turning point in the history of astronomy, it would be rather because of this enunciation than on account of the insufficient evidence of his astronomical discoveries (the triple motion of the earth).[66] One has, however, to keep in mind that Cusanus' principally mystic-speculative approach to his conclusions is fundamentally different from the scientific method of the Renaissance.

The theories of both Crescas and Cusanus, nevertheless, were far in advance of their time. If the notion of space was to be emancipated from the Aristotelian tradition, it would have to be done, as history proved, more gradually. It was not done until the sixteenth century. Even in Cardan's *De subtilitate*, space is still conceived in accord with Aristotelian tradition as the concave surface of the limiting body. "Est igitur locus ultima corporis superficies, corpus contentum ambicus." [67]

In contrast to Cardan, Scaliger identifies space with the void, which is coextensive with the body occupying it. Under the influence of atomistic thought, Scaliger presupposes the vacuum

[66] See Lynn Thorndike, *Science and thought in the fifteenth century* (New York, 1929), p. 133.
[67] Jerome Cardan, *De subtilitate*, lib. I.

as a necessary condition of motion. "In natura vacuum dari necesse est." [68] Scaliger's vacuum, however, is not an infinite empty extension beyond all bodies, but merely the receptacle coexistent with matter and penetrated by matter. The terms *"vacuum,"* *"locus,"* and *"spatium"* are synonymous in Scaliger's doctrine. "Idemque esse vacuum et locum; neque differre, nisi nomine." [69] Although Scaliger's theory represents an important step forward in the demolition of Aristotelian doctrine, it is still not the decisive step. For to Scaliger space, in its logical as well as metaphysical significance, is only secondary to matter. In a word, Scaliger's physics is still dominated by Aristotelian categories. As Ernst Cassirer points out, the real turning point is Bernardino Telesio's and Franciscus Patritius' theories of space. [70]

In his general philosophic outlook Telesio adopted certain materialistic and Stoic conceptions of Antiquity, which led him to ascribe to spiritual functions a certain degree of corporeality. This may account for his tendency to attribute independent reality to space and time, to place them on the same level with concrete matter. Space ceases with Telesio to be a mere quality and assumes an independent existence, parallel to matter or *"moles,"* *moles* being a concept that comes very near to the Newtonian notion of mass. Space is the great receptor of all being whatever. If a body leaves its place or is expelled from it, place itself does not leave, nor is it expelled, but remains the same, promptly becoming the receptacle of another body.

Itaque locus entium quorumvis receptor fieri queat et in existentibus entibus recedentibus expulsisve nihil ipse recedat expellaturve, sed idem perpetuo remaneat et succedentia entia promptissime suscipiat omnia, tantusque assidue ipse sit, quantaquae in ipso locantur sunt entia; perpetio nimirum iis, quiae in ea locata sunt, aequalis, at eorum nulli idem sit nec fiat unquam, sed penitus ab omnibus diversus sit. [71]

[68] J. C. Scaliger, *Exotericarum exercitationum liberi ad Hieronymum Cardanum* (Lutet, 1557).

[69] *Ibid.*

[70] E. Cassirer, *Das Erkenntnisproblem in der Philosophie und Wissenschaft der neueren Zeit* (Berlin, 1911).

[71] Telesio, *De natura rerum juxta propria principia libri novem* (Naples, 1586), I, 25.

Thus space, though equal to the things which occupy it, is not the same as any of these things. First of all, space is incorporeal, and, being pure aptitude to receive matter ("aptitudo ad corpora suscipienda"), it is free of all actions and operations. Space shows no qualitative differentiation; it is completely homogeneous in its structure, so that the existence of "natural places" is impossible. All parts of space show equal aptitude to receive any kind of matter. The motion of bodies in space is not caused by any qualitative differences inherent in space itself, but is the result of physical forces. Space as a whole is immobile ("universum perpetuo immobile permanet"). It is accessible to sense perception ("ipso comprehensum est sensu"), as experiments with vacua clearly show. Basing himself on physical grounds, Telesio attacks Aristotle's argument against the possibility of empty space, while disdaining to deal with demonstrations of the nonexistence of things whose existence is yet patently observable.

The considerations adduced by Telesio show clearly the new spirit of Italian natural philosophy of the sixteenth century. Nothing less than the formulation of a new physics is at issue. But another obstacle has still to be removed before these ideas could be assimilated and a new mechanics reared on their basis. The traditional substance-accident doctrine, the great bulwark of scholastic thought, had to be set aside. It was not enough to revise the physical foundations of the theory of space: it had to be provided with a new metaphysical foundation as well.

Franciscus Patritius undertook this task.

Quid ergo est? hypostasis, diastema, est, diastasis, ectasis est, extensio est, intervallum est, capedo est, atque intercapedo. Ergo quantitas? Ergo accidens? Ergo accidens ante substantiam? & ante corpus? Architas uterque, & senior Pythagorae auditor, & iunior Platonis amicus, & quicos secuti sunt scriptores categoriam, hoc spacium non cognovere.[72]

Is space a substance or an accident, is it corporeal or incorporeal, he asks in the chapter called "De spacio physico" of his com-

[72] Patritius, *Nova de universis philosophia libris quinquaginta comprehensa* (Venice, 1593), fol. 65.

prehensive work. None of these concepts applies to space, since they are only ways of characterizing things in space. Space must be presupposed as a necessary condition of all that exists in it. "Id enim ante omnia necesse est esse, quo posito alia poni possunt omnia; quo ablato alia omnia tollantur." [73] Further, qualities themselves are still dependent on space. It is therefore clear that space does not fit into the substance-accident scheme. "Nulla ergo categoriarum spatium complectitur; ante eas est, extra eas omnes est . . ." Patritius thus achieves the important result of emancipating the concept of space from the Aristotelian doctrine of categories. But, he asks, has not space magnitude? And is it not therefore subjected to the category of quantity? And this is his answer:

Itaque aliter de eo philosophandum est quam ex categoriis. Spatium ergo extensio est hypostatica per se substans, nulli inhaerens. Non est quantitas. Et si quantitas est, non est illa categoriarum, sed ante eam ejusque fons et origo.[74]

This view of space as being ontologically and epistemologically the primary basis of all existence led Patritius, as Cassirer points out,[75] to reverse the relation between mathematics and physics. The study of space must come before the study of matter. To Patritius, since space conditions not only matter as such, but its qualities as well, the investigation of space is an indispensable prerequisite to all natural science. Space makes not only nature, but the knowledge of nature, possible.

Before we go on to analyze Patritius' influence on the development of seventeenth-century physics, we may pause for a moment to say something about Giordano Bruno's place in the history of the development of the concept of space. As the exponent of the philosophy of infinity, Bruno is obliged to dispose of the idea of the world's finiteness, and he is thus confronted with the Peripatetic physics, in particular, with Aristotle's definition of place. "If the world is finite, and beyond the world there is

[73] Ibid.
[74] Patritius, Pancosmia. De spatio physico, 65 f.
[75] Cassirer, Das Erkenntnisproblem, vol. 1, p. 232.

nothing at all, where then is the world?" asks Bruno. Aristotle's answer that the world is in itself, although it follows logically from the definition of place, does not satisfy Bruno. So without attacking the validity of the logical conclusion Bruno confines himself to the premise itself. It is the definition itself that is wrong, and only a wrong conclusion could follow. To define place as the adjacent boundary of the containing body is to preclude the existence of space for the outermost sphere, and this renders meaningless any question as to what is outside the world. Before stating his own ideas, Bruno, in the manner of Crescas, mentions the arguments of Aristotle: "The convex surface of the primal heaven is universal space, which being the primal container is by naught contained. For position in space is no other than the surfaces and limit of the containing body, so that he who hath no containing body hath no position in space." [76] On the question "Where is the universe?" Aristotle, on the basis of these definitions, can only answer: "It is in itself." Here it is where Bruno's criticism begins. He says: "What then dost thou mean, O Aristotle, by this phrase, that 'space is within itself'? What will be thy conclusion concerning that which is beyond the world? If thou sayest, there is nothing, then the heaven and the world will certainly not be anywhere." After the discussion on the importance of the convex surface of the outermost sphere for spatial relations, Bruno (through the words of Philotheo) confesses: "Thus let this surface be what it will, I must always put the question, what is beyond?" [77] Bruno's restless temperament and his constantly searching disposition of mind did not let him find satisfaction with Peripatetic dialectic. Rejecting the finite categories of Peripatetic thought, he forms an ecstatic vision of an infinite universe in his mind.

> Henceforth I spread confident wings to space;
> I fear no barrier of crystal or of glass;
> I cleave the heavens and soar to the infinite.

[76] Bruno, *On the infinite universe and worlds*, trans. by Dorothea Waley Singer in *Giordano Bruno* (Schuman, New York, 1950), p. 251.
[77] *Ibid.*, p. 254.

And while I rise from my own globe to others
And penetrate ever further through the eternal field,
That which others saw from afar, I leave far behind me.[78]

It is therefore only natural that Bruno expresses a new conception of infinite space on the ground that "Si non superficies sed
spatium quoddam locus est, nullum corpus neque ulla corporis
illocata erit sive maximum, sive minimum sive finitum sit ipsum,
sive infinitum." [79] Bruno's definition of space is contained in Philotheo's answer on Albertino's theses in the fifth dialogue of *On
the infinite universe and worlds*. Replying to Albertino's fifth and
sixth arguments Philotheo says:

There is a single general space, a single vast immensity which we
may freely call VOID; in it are innumerable (*innumerabili et infiniti*)
globes like this one on which we live and grow. This space we declare
to be infinite, since neither reason, convenience, possibility, sense-
perception nor nature assign to it a limit . . . It diffuseth throughout
all, penetrateth all and it envelopeth, toucheth and is closely attached
to all, leaving nowhere any vacant space; unless, indeed, like many
others, thou preferest to give the name of void to this which is the
site and position of all motion, the space in which all have their
course.[80]

Although this definition, or description, of space is characteristic of the spirit of Italian natural philosophy of the sixteenth
century, it is yet the case, as Wolfson points out, that a certain
indebtedness of Bruno to Crescas is likely. Both Crescas and Bruno
focus their critique of Aristotle's definition on the problem of the
outermost sphere; both attempt to demonstrate the existence of
a vacuum on similar grounds; both refute Aristotle's theory of
lightness in much the same way.

That two men separated by time and space and language, but
studying the same problems with the intention of refuting Aristotle,
should happen to hit upon the same arguments is not intrinsically impossible, for all these arguments are based upon inherent weaknesses
in the Aristotelian system. But knowing as we do that a countryman

[78] *Ibid.*, p. 249.
[79] Bruno, *Acrotismus* (Vitebergae, 1588), I, 1, p. 121.
[80] Bruno, *On the infinite universe and worlds;* see reference 76, pp. 363,
373.

of Bruno, Giovanni Francesco Pico della Mirandola, similarly separated from Crescas in time and space and language, obtained knowledge of Crescas through some unknown Jewish intermediary, the possibility of a similar intermediary in the case of Bruno is not to be excluded.[81]

Campanella develops Patritius' theory of space still further, maintaining that space is the immovable basis of all existence: "basin omnis creati, omniaque praecedere esse saltem origine et natura." [82] At another place he calls space "locus, basis existentiae, in quo pulcrum Opificium, hoc est mundus, sedet." [83] In Campanella's view space is homogeneous and undifferentiated, penetrated corporeally and penetrating incorporeally. Its homogeneity excludes such differentiations as "down" or "up," which attach to the diversities of bodies, rather than to space. It goes without saying that the existence of "natural places" is emphatically rejected. God created space as a "capacity," a receptacle for bodies. "Locum dico substantiam primam incorpoream, immobilem, aptam ad receptandum omne corpus." [84]

The works of Telesio, Patritius, and Campanella show that Italian natural philosophy must be credited with having emancipated the concept of space from the scholastic substance-accident scheme. In the physics of the early seventeenth century space becomes the necessary substratum of all physical processes. It is this emancipated concept, divested of all inherent differentiations or forces. Gilbert in his *Philosophia nova* expresses these ideas in a concise way:

Sed non locus in natura quicquam potest: locus nihil est, non existit, vim non habet; potestas omnis in corporibus ipsis. Non enim Luna movetur, nec Mercurii, aut Veneris stella, propter locum aliquem in mundo, nec stellae fixae quietae manent propter locum.[85]

[81] Wolfson, *Crescas' critique of Aristotle*, p. 36.
[82] Thomas Campanella, *De sensu rerum* (1620), I, c. 12.
[83] Campanella, *Metaphysicarum rerum juxta propria dogmata* (1638), pars I, lib. 2, c. 13.
[84] Campanella, *Physiologia* (Paris, 1637), I, 2.
[85] William Gilbert, *De mundo nostro sublunari philosophia nova* (Amsterdam 1651), lib. II, cap. 8, p. 144.

Place does not affect the nature of things, it has no bearing on their being at rest or being in motion.

Although these words are directed first of all against the theory of "natural places" and of the attractive forces exerted by them, they mean much more. In order to understand the full meaning of "vim non habet," in which Gilbert agrees with Telesio, Patritius, and Campanella, we have to refer to an instrument which generally has been used since Antiquity for the measurement of *time,* but which paradoxically had a most important effect on the formation of concepts of *space:* the clepsydra. Since the Italian natural philosophers mention this device in their writings and since they draw important conclusions from the way it works, it is perhaps most fitting to explain the historical importance of the clepsydra for our subject in this chapter.

Aristotle,[86] referring to Anaxagoras, had stressed the importance of experiments with the clepsydra for the investigation into the existence of empty space. He, as well as his followers, quoted the rising of water in exhausted tubes, for instance, as a demonstration for the impossibility of a void. Philoponus, as we have seen,[87] denied the existence of a void "in actu," notwithstanding his criticism of his master's conception of space; and one of his arguments for his contention was just this kind of experiment.[88] A similar attitude was adopted by the author, or the authors, of the *Problemata,*[89] as well as by most of the other Peripatetics.

A different interpretation of the same phenomena is found in the writings of various Arabian authors. Biruni,[90] Avicenna,[91] and in particular al-Razi[92] not only cite these experiments as a verification of the existence of the void, but ascribe to empty

[86] Aristotle, *Physics,* IV, 6, 213 a.
[87] Page 54.
[88] See his commentaries to the *Physics,* 569.
[89] *Problemata,* XVI, 8 (trans. by E. S. Forster; Oxford, 1927).
[90] Biruni, *Al-ātār al-bāqiya,* p. 263; cf. C. E. Sachau, ed. and trans., *The chronology of ancient nations. An English version of the Arabic text of the Athār-ul-Bākiya of Albirūni* (London, 1879).
[91] Avicenna, *Sufficientia* (Venice, 1508), fol. 30 b.
[92] According to Fahr al-Dīn and Šīrazī.

space a force of attraction (ǧāḍiba).[93] Although al-Razi's conception of space is similar to that of Philoponus — with the important difference, of course, that al-Razi affirms the existence of
a void "in actu" — it may also be compared with the Stoic conception. For al-Razi assumes, as did the Stoa, two kinds of void,
the intramundane vacuities (ǧawhar al-ḫalāʾ or the "substance
of emptiness") and the void beyond the material universe (al-
faḍāʾ). Whether the force of attraction exerted by the void is
caused by a tendency of the intramundane vacuities to con-
glomerate and to unite among themselves, or whether it is to
be explained by their tendency to reach the faḍāʾ is hard to
ascertain. Both alternatives are conceived under the influence of
Platonism and both alternatives are instrumental for the explanation of the rising of light bodies, since lightness was explained
as a preponderance — not in the literal sense of the word — of
intramaterial vacua.

This conception of an empty space endowed with forces had
already been submitted to severe criticism by Roger Bacon[94]
and is now emphatically rejected by the Italian natural philosophers.[95]

Pierre Gassendi, who introduced his contemporaries to the
oldest complete source of atomism,[96] was particularly obliged to
face the problem of space and void. For as the proponent of a
revised atomism he was under the necessity of defending the
reality of the vacuum, which becomes in his view identical with
space. Although space and time appear to be "nothing," if measured by the scale of corporeal-concrete reality, they yet have real

[93] It is not impossible that the origin of these ideas lies in the following
simple experiment, cited by Heron in his *Pneumatics:* If you exhaust the
air from a small bottle with your mouth, the bottle remains attached to
your lips as if the void produced attracted your flesh. See H. Diels, "Ueber
das physikalische system des Straton," *Sitzber. preuss. Akad. Wiss. Berlin*
(1893), p. 101.

[94] P. Duhem, *Roger Bacon et l'horreur du vide* (Oxford, 1914).

[95] Telesio, *De natura rerum,* I, 25; Patritius, *Pancosmia,* I; Campanella,
De sensu rerum, I, 10.

[96] Pierre Gassendi, *Animadversiones in decimum librum Diogenis Laertii*
(Lyons, 1649).

existence, as the very preconditions of kinematics or physics in general. Further, Gassendi accepts Patritius' thesis of the priority of space over matter: "Ideo videntur Locus et Tempus non pendere a corporibus, corporeaque adeo accidentia non esse." [97] According to Gassendi, this priority is not only logical or ontological, but also temporal, for he says explicitly: "Unum est, spatia immensa fuisse, antequam Deus conderet mundum." Although the atoms were created by God, space was ever-existent, uncreated and independent. Gassendi was fully aware of the difficulties involved in this statement, which as we know was attacked later by Leibniz as an assertion of God's limitation. But Gassendi stresses the fact that famous theologians adhered to it. To Gassendi space is a necessary, infinite, immobile, and incorporeal datum of three dimensions. It is certainly no fiction, not even the mode of a substance. "Cum ex deductis constet posse quidem ea spatia dici nihil corporeum, seuquale substantia, aut accidens est, sed non nihil incorporeum ac specialis sui generis: constat quoque esse ea posse, etsi intellectus non cogitet, ac non quemadmodum chimaeram merum esse opus imaginationis." [98] Space is neither a mode nor an attribute; both of these exist in subordination to the object to which they belong, whereas space is independent of any substance. Bernier, the expounder of Gassendi's doctrine, in his *Abregé de la philosophie de Gassendi* [99] emphasizes the difference between Gassendi's space and the common notion of corporeal extension, and warns his readers not to confuse the two. Whereas space is infinite, corporeal extension is finite. Space can be occupied by bodies, but corporeal extension is impenetrable, subjected to all the vicissitudes of matter, whereas space is unchangeable and immovable.

It is certainly an important fact from our point of view that Gassendi and Campanella met personally. That Campanella's conception of a homogeneous and infinite space must have found

[97] Gassendi, *Syntagma philosophicum* (Florence, 1727), part II, sec. 1, lib. II, cap. 1.
[98] *Ibid.* I, 189.
[99] Second edition, Lyons, 1684, vol. 2, p. 9.

a ready support in Gassendi may also be seen in detail in Gassendi's *Epistolae tres de motu impresso a motore translato*.[100] In the atomism of Democritus and Epicurus he found the undifferentiated void through which the atoms move. From Gilbert and Kepler he adopted the idea that attraction and forces in general are not inherent to certain regions of space but have their causes in matter alone. The independence, autonomy, and priority of space, all vigorously propounded by Gassendi, were a timely concession to the requirements of the new physics. Physical phenomena could now be explained on the assumption of an infinite space that was partly filled and partly empty. Hence Gassendi's conception of space became the foundation, both of the atomistic theories of the seventeenth century with their discontinuous matter filling continuous space, on the small scale, and of celestial mechanics on the large scale. It was Newton who incorporated Gassendi's theory of space into his great synthesis and placed it as the concept of absolute space in the front line of physics.

[100] Gassendi, *Opera omnia* (Florence, 1727), vol. 3.

CHAPTER 4

THE CONCEPT OF

ABSOLUTE SPACE

Newton's conceptual scheme, as expounded in his *Philosophiae naturalis principia mathematica,* became the basis of classical physics and as such the subject of much profound analysis. We need mention only Neumann and Mach, who investigated its epistemological implications, and Wolff and Hegel, who explored its metaphysical foundations. So far as the purely physical teachings of the *Principia* are concerned, they are susceptible of different epistemological and metaphysical interpretations; for that work, as the first comprehensive hypothetico-deductive system of mechanics, lends itself, as does every system of the kind, to a variety of philosophical constructions. And so questions arise that allow of no absolute answer. Newton himself appears to have understood the distinction between the purely theoretical-deductive part of a theory and its practical application. In the Scholium to Proposition LXIX in the first book he says: "In mathematics we are to investigate the quantities of forces with their proportions consequent upon any conditions supposed; then, when we enter upon physics, we compare those proportions with the phenomena of Nature, that we may know what conditions of those forces answer to the several kinds of attractive bodies." [1] The comparison to which Newton here alludes (*conferendae sunt*)[2] seems to correspond to an "epistemic correlation" [3] in modern philosophy of science, except for Newton's quite different conception of the character of mathematics (*mathesis*). For to Newton mathematics, particularly geometry, is not a purely hypothetical system of propositions, logically deducible from axioms and definitions; instead geometry is nothing but a special branch of mechanics. "Therefore geometry is founded in mechanical practice, and is nothing but that part of universal mechanics which accurately proposes and demonstrates the art of measuring." [4]

This view of the relation of geometry to mechanics, Newton believes, follows from the impossibility of abstract geometry.

For the description of right lines and circles, upon which geometry is founded, belongs to mechanics. Geometry does not teach us to draw these lines, but requires them to be drawn, for it requires that the

[1] F. Cajori, ed., *Sir Isaac Newton's Mathematical principles of natural philosophy and his System of the world. A revision of Mott's translation* (University of California Press, Berkeley, 1934) [quoted as *Principles*], p. 192.

[2] For the original Latin text, references are given from the Thomson-Blackburn edition of the *Principia* (Glasgow, 1871) [quoted as *Principia*]. On p. 188 we read: ". . . deinde, ubi in physicam descenditur, conferendae sunt hae rationes cum phaenomenis . . ."

[3] Cf. F. S. C. Northrop, *The logic of the sciences and humanities* (New York, Macmillan, 1947), p. 119.

[4] Newton, *Principles*, p. xvii; *Principia*, p. xiii, "Auctoris Praefatio ad Lectorem," reads: "Fundatur igitur geometria in praxi mechanica, & nihil aliud est quam mechanicae universalis pars illa, quae artem mensurandi accurate proponit ac demonstrat."

learner should first be taught to describe these accurately before he
enters upon geometry, then it shows how by these operations problems
may be solved. To describe right lines and circles are problems, but
not geometrical problems. The solution of these problems is required
from mechanics, and by geometry the use of them, when so solved,
is shown.[5]

Newton's view of the unity of geometry and mechanics (cf.
his conception of "fluxions" and his aversion to handling geometric problems algebraically) can be traced back to his teacher
Isaac Barrow, for whom geometric curves had essentially a
mechanical character. In "De quadratura curvarum" Newton
writes: "Quantitates mathematicas, non ut ex partibus quam
minimis constantes, sed ut motu continuo descriptas, hic con-
sidero . . . Hae geneses in rerum nature locum vere habent et
in motu corporum quotidie cernuntur."[6] This realistic conception of mathematics is of the first importance for Newton's notion of absolute space, as we shall soon see. At this point it
interests us as being an important feature of Newton's methodology, showing, as it does, that the primary concepts underlying the structure of Newton's system are not hypothetical and
unreal, justified only by subsequent experimental verification.

It should be borne in mind, too, that such a remark applies
not only to the mathematical apparatus employed in the *Principia*, but to its fundamental laws, as for example the laws of
motion. We can see today that these laws are assumptions in-
accessible to experimental verification, but to Newton they were
facts of immediate experience. For although Newton calls the
laws of motion "axioms" (*Axiomata sive leges motus*), the term
"axiom" as employed by Newton in this context certainly does
not have the modern meaning of an arbitrary assumption; phrases
like "lex tertia . . . per theoriam comprobata est"[7] or "certa sit
lex tertia motus"[8] show clearly that Newton by his use of the
term axiom thought the relevant statement to be the point of

[5] Newton, *Principles*, p. xvii.
[6] *Opuscula Newtoni* (Lausanne and Geneva, 1744), vol. 1, p. 203.
[7] Newton, *Principia*, p. 25.
[8] *Ibid.*, p. 27.

departure for further investigation, and thus in conformity with his general plan, of which he writes: "To derive two or three general principles of motion from phaenomena, and afterwards to tell us how the properties and actions of all things follow from those manifest principles would be a very great step in philosophy." [9] It is in the light of these remarks that the historical treatment of Newton's theory of space must proceed. In other words, as historians we are bound to view Newton's system of mechanics not from the vantage point of a modern textbook on classical mechanics, but from that which Newton himself adopted. Accordingly we shall not confine ourselves to the *Principia* alone, but will take into consideration other writings of his as well, for example, the *Opticks,* the correspondence, and especially the famous exchange of letters between Leibniz and Newton's disciple, Samuel Clarke, who wrote under the guidance of the master.

Although Newton cannot, as we have already remarked, be regarded as a positivist in the modern sense of the word, yet he drew a clear line of demarcation between science on the one hand and metaphysics on the other. The famous "Hypotheses non fingo," although originally expressed only with relation to an explanation of gravitation, became his motto for the exclusion of the occult, metaphysical, or transcendental religious entities. His aim was not to abolish metaphysics, but to keep it distinct from physical investigation. It is well known that Newton, himself a religious man, never denied the existence of beings and entities that transcend human experience; he contended only that their existence had no relevance to scientific explanation: In its *mundus discorsi*, science has no place for them. Intimately acquainted with the problems of religion and metaphysics, Newton managed to keep them in a separate compartment of his mind, but for one exception, namely, his theory of space. Space thus occupies a unique place in his teachings.

In order fully to understand the Newtonian idea of space, it

[9] Newton, *Opticks* (ed. 4, London, 1730; Dover, New York, 1952), p. 401.

is necessary to bear in mind the general conceptual background of his physical system. Apart from space and time, force and mass are the fundamental concepts of the Newtonian physics. In Newton "force" is not the sophisticated notion of modern physics. It is not a mathematical abstraction, but an absolutely given entity, a real physical being. As for "mass," Newton, reverting to the view of Galileo, conceives of it as the most essential attribute of matter and thus places himself in diametrical opposition to Descartes, who identified matter with extension and regarded extension as the chief characteristic of matter. The Newtonian concept of "mass-point," still used in present-day textbooks, marks the chasm that separates Newton's concept of mass from Descartes' concept of spatial extension. A priori, it was perhaps a matter of predilection or preference which of the two, mass or extension, was to be given priority, since every real body has both and is inconceivable apart from either. Newton's abstraction proved to be the more fruitful.

Since mechanics deals with motion, space as the correlate of mass-point — just as the void was the correlate of the atom — has to be introduced at the very beginning of the system. It is therefore no accident that almost at the very beginning of the *Principia* we find the famous Scholium dealing with the concept of space.

I do not define time, space, place, and motion, as being well known to all. Only I must observe, that the common people conceive those quantities under no other notions but from the relation they bear to sensible objects. And thence arise certain prejudices, for the removing of which it will be convenient to distinguish them into absolute and relative, true and apparent, mathematical and common . . .
Absolute space in its own nature, without relation to anything external, remains always similar and immovable. Relative space is some movable dimension or measure of the absolute spaces; which our senses determine by its position to bodies; and which is commonly taken for immovable space; such is the dimension of a subterraneous, an aerial, or celestial space, determined by its position in respect to the earth. Absolute and relative space are the same in figure and magnitude; but they do not remain always numerically the same. For if the earth, for instance, moves, a space of our air, which relatively

and in respect of the earth remains always the same, will at one time be one part of the absolute space into which the air passes; at another time it will be another part of the same, and so, absolutely understood, it will be continually changed.[10]

In believing that time, space, place, and motion are concepts well known to all, Newton, as we see, does not feel called upon to give a rigorous and precise definition of these terms. Yet, because these notions arise only in connection with sensible objects, certain prejudices cling to them, and to overcome these Newton deemed it necessary to set up the distinctions of absolute and relative, true and apparent, mathematical and common. Since space is homogeneous and undifferentiated, its parts are imperceptible and indistinguishable to our senses, so that sensible measures have to be substituted for them. These coördinate systems, as they are called today, are Newton's relative spaces.

But because the parts of space cannot be seen, or distinguished from one another by our senses, therefore in their stead we use sensible measures of them. For from the positions and distances of things from any body considered as immovable, we define all places; and then with respect to such places, we estimate all motions, considering bodies as transferred from some of those places into others. And so, instead of absolute places and motions, we use relative ones; and that without any inconvenience in common affairs.[11]

In modern physics, coördinate systems are nothing but a useful fiction. Not so for Newton. Given Newton's realistic conception of mathematical objects, it is easy to understand why these relative spaces form "sensible measures." Not only is the reference body accessible to our senses, but likewise the "relative space" is dependent on it. But this accessibility to sense perception yields a notion that is of temporary validity only and lacking in generality. It is quite possible that there is no body at rest, to which the places and motions of other bodies may be referred; in a word: all these relative spaces may be moving coördinate systems. But moving in what? In order to answer this question, Newton takes flight from the realm of experience

[10] Newton, *Principles,* p. 6.
[11] *Ibid.,* p. 8.

altogether, at least for the time being. In his famous words, "But in philosophical disquisitions, we ought to abstract from our senses," [12] Newton introduces absolute and immutable space, of which relative space is only a measure. The final degree of accuracy, the ultimate truth, can be achieved only with reference to this absolute space. And it is therefore rightly called "true space."

What, it may be asked at this point, guarantees the final truth of absolute space, the very conception of which appears to contradict Newton's methodological rule: "We are to admit no more causes of natural things than such as are both true and sufficient to explain their appearances"? [13] In Newton's time this question became a highly controversial one and remained so until the beginning of the twentieth century. Is the concept of an absolute space a necessity for physics? Or can a consistent conceptual scheme be constructed that explains all physical phenomena without the use of such a concept? As every historian of physics knows, the problem reappeared in the nineteenth century as the problem of the ether and gave rise to an immense amount of discussion and experiment.

To Newton, absolute space is a logical and ontological necessity. For one thing, it is a necessary prerequisite for the validity of the first law of motion: "Every body continues in its state of rest, or of uniform motion in a right line, unless it is compelled to change that state by forces impressed upon it." [14] Rectilinear uniform motion requires a reference system different from that of any arbitrary relative space. Further, the state of rest presupposes such an absolute space. Newton explains:

Absolute motion is the translation of a body from one absolute place into another; and relative motion, the translation from one relative place into another. Thus in a ship under sail, the relative place of a body is that part of the ship which the body possesses; or that part of the cavity which the body fills, and which therefore moves together

[12] Ibid.
[13] Ibid., p. 398.
[14] Ibid., p. 13.

with the ship: and relative rest is the continuance of the body in the same part of the ship, or of its cavity. But real, absolute rest, is the continuance of the body in the same part of that immovable space, in which the ship itself, its cavity, and all that it contains, is moved. Wherefore, if the earth is really at rest, the body, which relatively rests in the ship, will really and absolutely move with the same velocity which the ship has on the earth. But if the earth also moves, the true and absolute motion of the body will arise, partly from the true motion of the earth, in immovable space, partly from the relative motion of the ship on the earth; and if the body moves also relatively in the ship, its true motion will arise, partly from the true motion of the earth, in immovable space, and partly from the relative motions as well of the ship on the earth, as of the body in the ship; and from these relative motions will arise the relative motion of the body on the earth.[15]

Since the first law of motion, as we have seen, is for Newton a matter of immediate experience, and since the law depends for its validity upon an absolute reference system, absolute space becomes indispensable to Newtonian mechanics. The interesting point, however, is that for Newton the introduction of the concept of absolute space into his system of physics did not result from methodological necessity only. Newton was led by his mathematical realism to endow this concept, as yet merely a mathematical structure, with independent ontological existence. He realized that there was a great difficulty to be overcome: the "inertial system," or, in less modern words, the system in which the first law holds, is not uniquely determined. Newton's mechanics is invariant for a translational transformation with constant velocity, that is, a Galilean transformation. Newton recognizes that a whole class of "spaces" or reference systems comply with this requirement. In Corollary V we read: "The motions of bodies included in a given space are the same among themselves, whether that space is at rest, or moves uniformly forwards in a right line without any circular motion." [16]

If Newton had been a confirmed positivist he would have acknowledged all uniformly moving inertial systems as equivalent

[15] *Ibid.*, p. 7.
[16] Ibid., p. 20.

to each other. As it was, only one absolute space existed for him. How is this space to be distinguished from among the multitude of inertial systems? For the solution of this problem Newton resorts to cosmology. In Hypothesis I of his *The system of the world* [17] he states: "That the centre of the system of the world is immovable. This is acknowledged by all, while some contend that the earth, others that the sun, is fixed in that centre."

To Newton, now, the center of the world is the center of gravity of the system composed of the sun, the earth, and the planets;[18] this center either is at rest or moves uniformly forward in a straight line; the latter alternative, however, is eliminated by Hypothesis I. In this way Newton defines the unique absolute space among all possible inertial frames. It it interesting to note that in the last-mentioned Corollary Newton is concerned to find the astronomical location of this universal center of gravity, which is his reference point for the determination of absolute space. He maintains that the movable centers of the earth, sun, and planets cannot serve as such a center, since they all gravitate toward each other. However, if the body toward which other bodies gravitate most has to be placed in the center, then it is the sun that should be allowed this privilege. Yet, since the sun itself is moving, a fixed point has to be chosen from which the center of the sun recedes least, and from which, if its density and volume were greater, it would recede still less.

All this points to the rather limited scope of Newton's cosmological conceptions. It is also interesting to note that Newton did not take into account the fixed stars when trying to determine the center of gravity of the world. Had he done so, he might have come very near the conception of the body "Alpha," which was introduced by C. Neumann[19] at the end of the last century. The fact that Newton ignored the fixed stars in this respect is the more curious, since for him they were still really "fixed," that is,

[17] *Ibid.*, p. 419.
[18] *Ibid.*, p. 419, corollary to Proposition XII.
[19] C. Neumann, *Ueber die Prinzipien der Galilei–Newton'schen Theorie* (1869).

not moving in space. For although Bruno had already imagined the sun to be in motion, and although Halley confirmed this anticipation in 1718, when he announced [20] that Sirius, Aldebaran, Betelgeuse, and Arcturus had unmistakably shifted their positions in the sky since Ptolemy assigned their places in his catalogue, it was only after the death of Newton that the proper motion of the stars became an accepted truth.

Newton's cosmological assumption that the center of the world is at rest escapes all possibility of experimental or observational verification. The fact was clearly recognized by Berkeley, one of the great opponents of the theory of absolute space. In "De motu" he writes: "Uti vel ex eo patet quod, quum secundam illorum principia qui motum absolutum inducunt, nullo symptomate scire liceat, utrum integra rerum compages quiescat, an moveatur uniformiter in directum, perspicuum sit motum absolutum nullius corporis cognosci posse." [21]

As we shall see in what follows, Newton was convinced that dynamically, though not kinematically, absolute space can be determined through the existence of centrifugal forces in rotational motion. Although Newton does not explicitly draw the conclusion that centrifugal forces determine absolute motion which in its turn determines absolute space, it is clear that this was his intention and it was always recognized as such by his commentators. If space is a physical reality, as Newton undoubtedly assumes, and if accelerated motion furnishes a criterion for its identification, it would appear to be a serious inconsequence to hold that uniform translational motion, since it fails to provide such a criterion, is different from all other kinds of motion; furthermore, space would seem to possess a dual structure, absolute for accelerated motion and relative for uniform translation. Newton's cosmological assumption protects

[20] E. Halley, *Phil. Trans.* 30, 737 (1718).

[21] A. A. Luce and T. E. Jessop, ed., *The works of George Berkeley* (Nelson, London, 1951), vol. 4, p. 28.

him against such an objection, which incidentally was raised by Leibniz in his correspondence with Huygens.

According to Newton, as we have seen, the first law of motion presumes the necessary existence of absolute space but provides no means by which it can be identified experimentally. Hence Newton's next step. Since absolute space and time "do by no means come under the observation of our senses," it becomes necessary to investigate the dynamics of motion. For motion, accelerated motion in particular, is the means and medium through which space can be explored. Inasmuch as they refer to relative or to absolute space, motions are either relative or absolute, so that if it were possible to identify absolute motion, the identification of absolute space would follow. Now absolute motion, according to Newton, can be distinguished from relative motion by its "properties, causes, and effects."

The causes by which true and relative motions are distinguished one from the other, are the forces impressed upon bodies to generate motion. True motion is neither generated nor altered, but by some force impressed upon the body moved; but relative motion may be generated or altered without any force impressed upon the body. For it is sufficient only to impress some force on other bodies with which the former is compared, that by their giving way, that relation may be changed, in which the relative rest or motion of this other body did consist . . .

The effects which distinguish absolute from relative motion are, the forces of receding from the axis of circular motion. For there are no such forces in a circular motion purely relative, but in a true and absolute circular motion, they are greater or less, according to the quantity of the motion . . .

It is indeed a matter of great difficulty to discover, and effectually to distinguish, the true motions of particular bodies from the apparent; because the parts of that immovable space, in which those motions are performed, do by no means come under the observation of our senses. Yet the thing is not altogether desperate; for we have some arguments to guide us, partly from the apparent motions, which are the differences of the true motions; partly from the forces, which are the causes and effects of the true motions.[22]

[22] Newton, *Principles*, pp. 10, 12.

Thus Newton's first argument with regard to absolute motion is based on the idea that real force creates real motion. To Newton, at least in this context, forces are metaphysical entities conceived anthropomorphically. However, if we leave out of account the import of forces for the determination of absolute space, the notion of force in Newton's mechanics may be interpreted in the modern functional way, as in Heinrich Hertz's *Die Prinzipien der Mechanik:* "Was wir gewohnt sind als Kraft und als Energie zu bezeichnen ist dann fuer uns nichts weiter als eine Wirkung von Masse und Bewegung, nur braucht es nicht immer die Wirkung grobsinnlich nachweisbarer Masse und grobsinnlich nachweisbarer Bewegung zu sein." [23]

But undoubtedly there is no question of this functional conception of force in Newton's discussion of absolute space. It is foreign to the general character of his system. His argument "from causes" is based on the traditional metaphysics, the inclusion of which in the framework of physical explanation is strongly objected to by Newton himself. In order to see the vicious circle inherent in Newton's reasoning, we have only to think for a moment of a world of moving masses in which no living organism existed. For in such a world an absolute force could be determined, according to Newton, solely by the absolute motion of the body on which this force was exerted.

The second argument for the existence of absolute motion proceeds from the effects that such motion produces, in particular, the appearance of centrifugal forces ("vires recedendi ab axe motus circularis"). So we have Newton's famous pail experiment, which he describes as follows:

If a vessel, hung by a long cord, is so often turned about that the cord is strongly twisted, then filled with water, and held at rest together with the water; thereupon, by the sudden action of another force, it is whirled about the contrary way, and while the cord is untwisting itself, the vessel continues for some time in this motion; the surface of the water will at first be plain, as before the vessel began to move; but after that, the vessel, by gradually communicating its

[23] H. Hertz, *Die Prinzipien der Mechanik* (Leipzig, 1894), p. 31.

motion to the water, will make it begin sensibly to revolve, and recede
by little and little from the middle, and ascend to the sides of the
vessel, forming itself into a concave figure (as I have experienced),
and the swifter the motion becomes, the higher will the water rise, till
at last, performing its revolutions in the same times with the vessel,
it becomes relatively at rest in it. This ascent of the water shows its
endeavor to recede from the axis of its motion; and the true and absolute circular motion of the water, which is here directly contrary to
the relative, becomes known, and may be measured by this endeavor.
At first, when the relative motion of the water in the vessel was greatest, it produced no endeavor to recede from the axis; the water showed
no tendency to the circumference, nor any ascent towards the sides
of the vessel, but remained of a plain surface, and therefore its true
circular motion had not yet begun. But afterwards, when the relative
motion of the water had decreased, the ascent thereof towards the
sides of the vessel proved its endeavor to recede from the axis; and this
endeavor showed the real circular motion of the water continually
increasing, till it had acquired its greatest quantity, when the water
rested relatively in the vessel. And therefore this endeavor does not
depend upon any translation of the water in respect of the ambient
bodies, nor can true circular motion be defined by such translation.[24]

For a clear analysis of this experiment, let us consider also
the final phase — not described by Newton — when the rotation of the pail is stopped while the water continues its circular
motion (owing to conservation of angular momentum). During
this final stage of the experiment, as long as friction can be ignored, the water contained in the vessel maintains its paraboloidal surface.

The gist of this experiment can be summarized in modern
terms as follows: Both in the beginning of the experiment (when
the pail spins alone) and at the end of the experiment (when the
water spins alone) pail and water are moving relative to each
other in the same manner. Rigorously considered, the directions
of the relative rotations are reversed; but owing to the assumed
isotropy of space this reversal can obviously have no effect on
the dynamical result. If in the second case the time parameter
had been reversed, as is permissible in a purely mechanical
phenomenon, exactly the same relative motion would have re-

[24] Newton, *Principles*, p. 10.

sulted. Now, were all motion (rotation) purely relative, no physical difference should become apparent between the two states. However, since the surface of the water contained in the pail is level in the first case and paraboloidal in the second, rotation, thus concludes Newton, must be absolute.

This experiment was the cause of much controversy in the history of modern physics and the situation was clarified only with the appearance of Einstein's principle of equivalence in his general theory of relativity. In Newton's interpretation of the pail experiment he obviously again transcends the realm of experience. His simple assumption that the surface of water in the pail would be as curved, even if it were rotating in empty space, as when rotating in space filled with starry matter, is not susceptible of physical verification. And the same inaccessibility to physical verification characterizes all the other attempts to inforce his argument, as, for example, his experiment with the two cord-connected spheres revolving around their common center of gravity, the tension in the cord being taken by him as an indication of the absolute motion of the spheres. "And thus we might find both the quantity and the determination of this circular motion, even in an immense vacuum, where there was nothing external or sensible with which the globes could be compared." [25] But such conditions can never be realized, any more than in the case of the astronomical effects of centrifugal forces, as for example the spheroidicity of the earth and of Jupiter, as Newton expounds the matter in the third book of his *Principia*.[26]

Berkeley rejects Newton's implicit assumption that the pail experiment, if performed in empty space, would yield the same result. As Berkeley explains in his "De motu," the real motion of the pail is far from being circular, if the diurnal rotation of the earth and its annual revolution are taken into account. For

[25] *Ibid.*, p. 12.
[26] Proposition XVIII, Theorem XVI; also Proposition XIX. Problem III (*Principles*, p. 424); *et alia*.

Berkeley the idea of an absolute motion and an absolute space is a mere fiction, lacking all experimental foundation. Relating all such motions as are exhibited in the pail experiment ultimately to the system of stars as frame of reference, Berkeley assumes the existence of such a system as necessary for the intelligibility of motion. He says:

> If we suppose the other bodies were annihilated and, for example, a globe were to exist alone, no motion could be conceived in it; so necessary is it that another body should be given by whose situation the motion should be understood to be determined. The truth of this opinion will be very clearly seen if we shall have carried out thoroughly the supposed annihilation of all bodies, our own and that of others, except that solitary globe.
> Then let two globes be conceived to exist and nothing corporeal besides them. Let forces then be conceived to be applied in some way; whatever we may understand by the application of forces, a circular motion of the two globes round a common centre cannot be conceived by the imagination. Then let us suppose that the sky of the fixed stars is created; suddenly from the conception of the approach of the globes to different parts of that sky the motion will be conceived.[27]

Berkeley's statement obviously cannot be considered as being equivalent to what is called in modern cosmology "Mach's principle" (that is, that the inertia of any body is determined by the masses of the universe and their distribution), as Berkeley confines himself to the problem of the perception and comprehensibility of motion and ignores in this context the dynamical aspect of motion.

Newton's final argument, based upon the distinction between absolute and relative, or apparent, motion, is not further developed in his works. His idea seems to be that a body which moves in relative motion may be moving in absolute motion or may be at rest relative to absolute space; there is no way of deciding between these two alternatives. However, if a second body moves relative to the first, it is clear that one of these two at least must be endowed with absolute motion. It is impossible that both of

[27] Luce and Jessop, ed., *The works of George Berkeley*, vol. 4, p. 47.

them should be at rest with respect to absolute space.[28] The weakness of this argument is its indefensible assumption that an absolute reference system is an essential prerequisite for the description of the behavior of these bodies.

So Newton takes over from Patritius, Campanella, and Gassendi the concept of an infinite space, which is homogeneous and isotropic and, in addition, succeeds in convincing himself that he has proved the reality of this concept by physical experiment. He thought he had demonstrated that space has an existence proper to itself and independent of the bodies that it contains. In his view, it makes sense, therefore, to state that any definite body occupies just *this* part of space and not another part of space, and the meaning of such a statement does not presuppose a relation to any other bodies in the universe. He was not aware that his procedure violated the very principles of the methodology he professed. Since he was a younger contemporary of Henry More, whose personal acquaintance he made in his youth and whose teachings, via Isaac Barrow, exerted a great influence upon him, it is no wonder that Newton found support for his theory of space in the doctrine of that thinker. More's important works had been published about seven years before the appearance of the *Principia*. But it was the religious element, originating, as we saw, in Jewish cabalistic and Neoplatonic thought, that gained ascendency over Newton in his later years. So a comparison of the first and later editions of the *Principia* shows that the identification of absolute space with God, or with one of his attributes, came into the foreground of Newton's thought only toward the end of his life, that is, at the beginning of the eighteenth century. However, his interest in Biblical and post-Biblical literature may be traced back to the influence of one of the teachers in Cambridge, Joseph Mede, a fellow of Christ's College. Mede, apart from his studies in apocryphical and other esoteric literature,

[28] This argument in defense of absolute motion reappeared later in Alois Hoefler's *Studien zur gegenwärtigen Philosophie der Mechanik* (Leipzig, 1900), p. 133.

stimulated philological interest among his students in the Hebrew of the Bible by his etymological theory, quite popular at that time, that Hebrew was the mother of all languages.

It also has been established that Durand Hotham's book on Jacob Böhme exerted a strong influence on young Newton. Böhme's *Mysterium magnum*, a commentary on Genesis, shows extraordinary parallels to the *Zohar* and to other sources of Jewish theosophy. The Hebrew *Chokmah*, a body of books ascribed to King Solomon, seemed to have passed over to the Gnostic *Sophia* and by another transition to the "Virgin Wisdom" of Böhme. We know also with certainty that Henry More[29] and Isaac Barrow exerted a very strong influence on Newton at that time. Henry More was the spiritual leader at Christ's College in Cambridge and the chief disseminator of Cabalistic and Neo-platonic ideas, as described in detail in Chapter II. Isaac Barrow, Newton's famous teacher, promulgated More's ideas in mathe-matized form in his *Mathematical lectures*. In Barrow's geometry, space is the expression of divine omnipresence, just as time is the expression of the eternity of God. Under the influence of these strong forces it seems most probable that Newton, even when writing on purely physical problems, had similar ideas in the back of his mind. In fact, that he had theological and religious ideas in his mind when writing the *Principia* is evident from his letter (December 10, 1692) to Richard Bentley in which he confessed: "When I wrote my treatise about our system, I had an eye upon such principles as might work with considering men for the belief of a Deity; and nothing can rejoice me more than to find it useful for that purpose." It was, however, only in 1713 that Newton prepared the General Scholium of Book III to be published in the second edition (1713). It is in this Scholium, in addition to Queries 19–31 of the *Opticks* (missing in the first edition), that we find explicit statements of Newton's ideas on the relation between his theory of absolute space and

[29] For the facts of personal contact between More and Newton, see L. T. More, *Isaac Newton* (Scribner, New York, 1934), pp. 11, 31, 182.

theology. Undoubtedly, Newton's increasing interest in theological and spiritual questions during his later years was one of the motives for the preparation of the Scholium. Another reason was Cotes's request that he prevent any recurrence of criticisms which pronounced Newton's theory of space as leading to atheism. In a letter (March 18, 1713) to Newton, the editor of the second edition of the *Principia* writes: "I think it will be proper to add something by which your Book may be cleared from some prejudices which have been industriously laid against it . . . That You may not think unnecessary to answer such Objections You may be pleased to consult a Weekly Paper called *'Memoires of Literature'* and sold by Ann Baldwin in Warwick-Lane." The article referred to is Leibniz's letter (February 10, 1711) to the Dutch physician Hartsoeker, in which Leibniz attacks Newton's theory of gravitation. Of greater relevance for our subject, however, is Berkeley's attack on Newton's theory of space, which Cotes certainly had in mind, although he did not mention Berkeley by name. Berkeley published in 1710 his *Principles of human knowledge,* in which he criticizes Newton's concept of absolute space on theological grounds as being a pernicious and absurd notion. Space, according to Berkeley, has to be conceived as relative only, "Or else there is something beside God which is eternal, uncreated, infinite, indivisible, unmutable." [30]

It is therefore not surprising that in the General Scholium Newton gives free reign to his religious enthusiasm:

It is the dominion of a spiritual being which constitutes a God: a true, supreme, or imaginary dominion makes a true, supreme, or imaginary God. And from true dominion it follows that the true God is a living, intelligent, and powerful Being; and, from his other perfections, that he is supreme, or most perfect. He is eternal and infinite; omnipotent and omniscient; that is, his duration reaches from eternity to eternity; his presence from infinity to infinity; he governs all things, and knows all things that are or can be done. He is not eternity and infinity, but eternal and infinite; he is not duration or space, but he

[30] Berkeley, *Principles of human knowledge,* in *A new theory of vision and other writings* (Dent, London, 1938), p. 173.

endures and is present. He endures for ever, and is everywhere present; and by existing always and everywhere, he constitutes duration and space.[31]

Here, for the time, Newton identifies space and time with God's attributes. God is not eternity and infinity, but he is eternal and infinite. Eternal and omnipresent, God constitutes duration and space. A few lines further on we read, "In ipso continentur & moventur universa," "In him are all things contained and moved," a statement to which Newton adds the marginal remarks that this was the opinion of the Ancients: St. Paul (Acts 17:27, 28), St. John (14:2), Moses (Deut. 4:39), David (Ps. 139:7–9), Solomon (I Kings 8:27), Job (22:12–14), Jeremiah (23:23, 24). Here we have unmistakably an echo of More's *Enchiridion metaphysicum* and his *Divine dialogues,* but with this difference, that Newton's expressions are more reserved and more carefully chosen. He seemed to be aware that he might easily be misunderstood and counted among the pantheistic thinkers of his time, who in orthodox circles were identified with the atheists.

Since every particle of space is always in existence, and every indivisible moment of duration is everywhere, "certainly the Maker and Lord of all things cannot be never and nowhere." [32] Elsewhere Newton speaks of

the Wisdom and Skill of a powerful ever-living Agent; who, being in all Places, is more able by his Will to move the Bodies . . . within his boundless uniform Sensorium, and thereby to form and reform the Parts of the Universe, than we are by our Will to move the Parts of our own Bodies.[33]

This identification of the omnipresence of space with the omnipresence of God leads to a serious difficulty, and Leibniz with his sharp intellect exploited it remarkably in his controversy with Clarke. For according to Newton's conception, the divisibility of space — relative spaces are parts of the absolute space — would

---

[31] Newton, *Principles,* p. 544.
[32] Newton, *Principia,* p. 528: "Certe rerum omnium fabricator ac dominus non erit nunquam, nusquam."
[33] Newton, *Opticks* (Dover ed.), p. 403.

appear to involve the divisibility of God. Clarke's response to Leibniz's argument may be summarized as follows: Absolute space is one; it is infinite and essentially indivisible. The assumption that it can be divided leads to a contradiction, since any partition — according to Clarke — would require an intermediary space. Hence divine infinity and omnipresence imply no divisibility of the substance of God. Clarke concludes that it is only because a pictorial and unjustified meaning is attached to the notion of divisibility that the difficulty arose.

Another point of interest in this controversy is the term "sensorium," occurring in the above quotation, and earlier in Query 28:

> . . . does it not appear from Phaenomena that there is a Being incorporeal, living, intelligent, omnipresent, who in infinite Space, as it were in his Sensory, sees the things themselves intimately, and throughly perceives them, and comprehends them wholly by their immediate presence to himself.[34]

In the letter opening the controversy, which was to end only with Leibniz's death in 1716, Leibniz says:

> Sir Isaac Newton says, that Space is an Organ, which God makes use of to perceive Things by. But if God stands in need of any Organ to perceive Things by, it will follow, that they do not depend altogether upon him, nor were produced by him.[35]

But did Newton really identify space with an organ of God? Or was this expression only an unfortunate *lapsus calmi*? Clarke's response to Leibniz gives the answer to this question:

> Sir Isaac Newton doth not say, that Space is the Organ which God makes use of to perceive Things by; nor that he has need of any Medium at all, whereby to perceive Things; But on the contrary, that he, being Omnipresent, perceives all Things by his immediate Presence to them, in all Space whereever they are, without the Intervention or Assistance of any Organ or Medium whatsoever. In order to make this more intelligible, he illustrates it by a Similitude: That as the Mind of Man, by its immediate Presence to the Pictures of Things,

---

[34] *Ibid.*, p. 370.
[35] *A Collection of Papers which passed between the late learned Mr. Leibnitz and Dr. Clarke* (London, 1717), p. 3.

form'd in the Brain by the means of the Organ of Sensation, sees those Pictures as if they were the Things themselves; so God sees all Things, by his immediate Presence to them: he being actually present to the Things themselves, to all Things in the Universe; as the Mind of Man is present to all the Pictures of Things formed in his Brain . . . And this Similitude is all that he means, when he supposes Infinite Space to be (as it were) the Sensorium of the Omnipresent Being.[36]

Accordingly, it seems to be clear that Newton used the term "Sensorium" merely as a comparison and did not identify space with an organ of God.

With Newton's conception of space now before us we may turn to the question why he thought it needful and appropriate to introduce theological considerations into the very body of his scientific writings. Apart from the reasons dictated by polemic, as we have seen, there are certainly other motives; John Tull Baker, in his monograph entitled *An historical and critical examination of English space and time theories,* discusses some of them. He writes that, in the first place, absolute space and time find a place in the *Principia* because as attributes of God they rendered to the *Principia* a completeness, as a cosmology, which it might have lacked otherwise. Furthermore, their inclusion in the very beginning of the Newtonian system gives to the foundations of mechanics and mathematical physics a theological justification, an idea congenial to Newton:

In the second place, the postulations of absolute time and absolute space suggest the construction of mathematical entities which might be approached as limits of perfection on the description of physical facts. Just as relative time always more nearly approaches absolute time as we refine our measurements and relative motion approximates absolute motion as we examine forces more carefully, so the scheme of things as a whole may be more clearly understood as we progress in more detailed experiment and analysis.[37]

According to this interpretation, the use of absolutes by Newton may be understood as an ideal of perfection, an ideal attain-

[36] *Ibid.,* p. 11.
[37] J. T. Baker, *An historical and critical examination of English space and time theories* (Sarah Lawrence College, Bronxville, New York, 1930), p. 30.

able in matters of space only. In addition it may be rightly claimed that absolute space and absolute time have always had a strong appeal to human emotion. Through their presence clarity and rigor, certainty and definiteness seem to be guaranteed.

One thing is certain: Newton's mechanics, as expounded in the *Principia,* is one great vindication of his theory of absolute space and absolute motion. At the end of the Scholium in the first book he says: "How we are to obtain the true motions from their causes, effects, and apparent differences, and the converse, shall be explained more at large in the following treatise. For to this end it was that I composed it." [38] "Hunc enim in finem tractatum sequentem composui." [39] To demonstrate the existence of true motion and absolute space — such is the program of the *Principia.* All Newton's achievements and discoveries in the realm of physics are in his view subordinate to the philosophical conception of absolute space. The outstanding success of Newtonian mechanics in the physics and astronomy of the last two centuries seemed an indubitable guarantee of the soundness of its philosophical implications. It is not surprising, therefore, that the criticisms leveled by Leibniz and Huygens against the theory of absolute space found no echo in this long period. Today we are in a position to understand the force of these criticisms, which is not to say that the *Principia* ceases to be a landmark in the history of human intellectual achievements. It is this not because of its philosophical conclusions but because of the wealth of its purely physical contents, backed by experimentation and hence verifiable, and further because of the wonderful systematization of this wealth of material.

It is not the purpose of this chapter to provide a comprehensive account of Leibniz's theory of space. In any case it is a task immensely complicated by the fact that Leibniz's theory in the course of its development passed through three differ-

[38] Newton, *Principles,* p. 12.
[39] Newton, *Principia,* p. 12.

ent stages at least. We shall confine our discussion here to his critique of Newton's conception, for the understanding of which it is necessary to bear in mind that in his view space is nothing but a system of relations, devoid of metaphysical or ontological existence. In his fifth letter to Clarke, Leibniz summarizes his conception of space as follows:

I will here show, how Men come to form to themselves the Notion of Space. They consider that many things exist at once, and they observe in them a certain Order of Co-existence, according to which the relation of one thing to another is more or less simple. This Order is their Situation or Distance. When it happens that one of those Co-existent Things changes its Relation to a Multitude of others, which do not change their Relation among themselves; and that another Thing, newly come, acquires the same Relation to the others, as the former had; we then say it is come into the Place of the former; and this Change we call Motion in That Body, wherein it is the immediate Cause of the Change. And though Many, or even All the Co-existing Things, should change according to certain known Rules of Direction and Swiftness; yet one may always determine the Relation of Situation, which every Co-existent acquires with respect to every other Co-existent; and even That Relation, which any other Co-existent would have to this, or which this would have to any other, if it had not changed or if it had changed any otherwise. And supposing, or feigning, that among those Co-existents, there is a sufficient Number of them, which have undergone no Change; then we may say, that Those which have such a Relation to those fixed Existents, as Others had to them before, have now the same Place which those others had. And That which comprehends all those Places, is called Space.[40]

Leibniz goes on to explain that the relation of situation is a wholly sufficient condition for the idea of space. No absolute reality need be invoked. He makes his point clear by an excellent illustration from genealogy:

In like manner, as the Mind can fancy to itself an Order made up of Genealogical Lines, whose Bigness would consist only in the Number of Generations, wherein every Person would have his Place: and if to this one should add the Fiction of a Metempsychosis, and bring in the same Human Souls again; the Persons in those Lines might change Place; he who was a Father, or a Grand-father, might become

[40] A Collection of Papers . . . , p. 195.

a Son, or a Grand-son &c. And yet those Genealogical Places, Lines, and Spaces, though they should express real Truths, would only be Ideal Things.[41]

The illustration of a tree of genealogy, which shows the mutual relations of kinship between certain persons by attributing to them definite positions within the scheme, serves Leibniz very well. For nobody would hypostatize this system of relations and endow it with ontological existence. Newton's absolute space, in Leibniz's view, is nothing but a similar unjustified hypostatization.

Having thus outlined his concept of space, Leibniz realizes that what he has done is only to define the expression "having the same place," this being enough for the foundation of the concept of physical space. He then proceeds with great ardor to attack More and through him Newton. For the context the following words of Leibniz are worth quoting:

If the Space (which the Author fancies) void of all Bodies, is not altogether empty; what is it then full of? Is it full of extended Spirits perhaps, or immaterial Substances, capable of extending and contracting themselves; which move therein, and penetrate each other without any Inconveniency, as the Shadows of two Bodies penetrate one another upon the Surface of a Wall? Methinks I see the revival of the odd Imaginations of Dr. Henry More (otherwise a Learned and well-meaning Man), and of some Others, who fancied that those Spirits can make themselves impenetrable whenever they please. Nay, some have fancied, that Man in the State of Innocency, had also the Gift of Penetration; and that he became Solid, Opaque, and Impenetrable by his Fall. Is it not overthrowing our Notions of Things, to make God have Parts, to make Spirits have Extension? [42]

Leibniz's clear conception of space[43] as a system of relations and his well-known "principium identitatis indiscernibilium" are the two solid foundations from which he launches his criticism of Newton's absolute space and absolute motion. On

[41] Ibid., p. 201.
[42] Ibid., p. 205.
[43] For a genetic history of Leibniz's philosophy of space and time, see W. Gent, "Leibnizens Philosophie der Zeit und des Raumes," Kantstudien 31, 61 (1926).

kinematic grounds there can be no doubt that Leibniz is the
victor in this dispute. Clarke's refutations of Leibniz's arguments
are often not to the point and show a great deal of misun-
derstanding. However, as soon as Clarke leaves the subject of
kinematics and brings forth — no doubt under the briefings
of Newton himself — the dynamical arguments in favor of the
existence of absolute space and motion, Leibniz faces an insuper-
able difficulty. With regard to Clarke's reference to the Scholium
and Newton's demonstrations therein of the existence of absolute
space and absolute motion by means of centrifugal forces, Leibniz feels obliged to admit:

However, I grant there is a difference between an absolute true
motion of a Body, and a mere relative Change of its Situation with
respect to another Body. For when the immediate Cause of the
Change is in the Body, That Body is truly in Motion; and then the
Situation of other Bodies, with respect to it, will be changed consequently, though the Cause of that Change be not in Them.[44]

Having thus bowed to the idea of an "absolute true motion,"
Leibniz is placed in a dilemma from which he finally sees only
one way out: namely, to allow for a double meaning of the concept of motion. On the one hand, it may denote the purely spatial change of situation, which saves his view of the conceptual
structure of space; on the other hand, it may signify a dynamical
process which is completely unrelated to space as such. But Leibniz is aware that such a stratagem exposes him to the danger
of having to fall back on doubtful scholastic concepts like
quality, form, substance. It is especially clear from Leibniz's
correspondence with Huygens that he tried desperately for years
without success to find a dynamical argument for the relativity
of motion. Yet it is a curious fact for us today to note that actually
he came very near to Mach's solution of the problem. In his
"De Causa Gravitatis, et Defensio Sententiae Autoris de veris
Naturae Legibus contra Cartesianos" [45] Leibniz tried to demonstrate that gravity is not explicable as a force acting at a distance,

[44] A Collection of Papers . . . , p. 213.
[45] Acta Eruditorum (1690).

but is reducible to the contiguous action of the surrounding ether. In other words, he tried to reduce gravity to a centrifugal force, saying: "Etsi valde dudum inclinaverim ipse ad gravitatem a vi centrifuga materiae aethereae circulantis repetendam, sunt tamen aliqua quae dubitationes gravissimas injecere." [46] Directly opposite to this was Mach's daring description of centrifugal forces as a disguised gravitational action. So Leibniz, having failed to find the key to dynamical relativity, saw no need to revise what he had written some twenty years before, when he summarized his remarks on Cartesian physics in his "Animadversions on Descartes' *Principles of Philosophy*":[47]

On Art. 25. If motion is nothing but change of contact or immediate vicinity, it follows that it can never be determined which thing is moved. For as in astronomy the same phenomena are presented in different hypotheses, so it is always permissible to ascribe real motion to either one or other of those bodies which change among themselves vicinity or situation; so that one of these bodies being arbitrarily chosen as if at rest, or for a given reason moving in a given line, it may be geometrically determined what motion or rest must be ascribed to the others so that the given phenomena may appear. Hence if there is nothing in motion but this respective change, it follows that no reason is given in nature why motion must be ascribed to one rather than to others. The consequence of this will be that there is no real motion. Therefore in order that a thing can be said to be moved, we require not only its situation in respect to others, but also that the cause of change, the force or action, be in itself.[48]

It is to these lines that Huygens refers in his letter of May 29, 1694 to Leibniz. He objects to the assertion "that it would be absurd, if there exists no real, but only relative motion" ("absonum esse nullum dari motum realem sed tantum relativum"). If Huygens' quotation from Leibniz is verbally inaccurate, it is not so essentially. Huygens declares his intention to stick to his theory — perhaps by way of contrasting his own firmness with Leibniz's wavering — and says that he will not let himself be

[46] G. I. Gerhardt, *Leibnizens mathematische Schriften* (Halle, 1860), part 2, vol. 6, p. 197.
[47] Published in 1692.
[48] G. M. Duncan, *The philosophical works of Leibnitz* (New Haven, 1890), p. 60.

influenced by the experiments in the *Principia*, convinced as he is that Newton is wrong. At the same time he hopes that Newton will retract his theory in the forthcoming second edition of the *Principia*, which he thought would be edited by David Gregory. Huygens' instinct toward his own theory was sound, although he was mistaken about the second edition of the *Principia*, which in fact was prepared by Roger Cotes, as he was mistaken about its possible revision by Newton.

The subject occurs in Huygens' first letter to Leibniz, which reads:

Je vous diray seulement, que dans vos notes sur des Cartes j'ay remarqué que vous croiez absonum esse nullum dari motum realem, sed tantum relativum. Ce que pourtant je tiens pour tres constant, sans m'arrester au raisonnement et experiences de Newton dans ses Principes de Philosophie, que je scay estre dans l'erreur, et j'ay envie de voir s'il ne se retractera pas dans la nouvelle edition de ce livre, que doit procurer David Gregorius.[49]

Leibniz's reply to this letter (June 22, 1694) is extremely interesting:

Quant à la difference entre le mouuement absolu et relatif, je croy que si le mouuement ou plus tost la force mouuante des corps est quelque chose de reel comme il semble qu'on doit reconnoistre, il faudra bien qu'elle ait un subjectum. Car a et b allant l'un contre l'autre, j'avoue que tous les phenomenes arriveront tout le meme, quel que soit celuy dans le quel on posera le mouuement ou le repos; et quand il y auroit 1000 corps, je demeure d'accord que les phenomenes ne nous scauroient fournir (ny même aux anges) une raison infallible pour determiner le sujet du mouuement ou de son degré; et que chacun pourroit estre conçû à part comme estant en repos, et c'est aussi tout ce que je crois que vous demandes; mais vous ne nieres pas je crois que veritablement chacun a un certain degré de mouuement on, si vous voulés de la force; non-obstant l'equivalence des Hijpotheses. Il est vray que j'en tire cette consequence qu'il y a dans la nature quelque autre chose que ce que la Geometrie y peut determiner. Et parmy plusieurs raisons dont je me sers pour prouuer qu'outre l'etendue et ses variations, qui sont des choses purement Geometriques, il faut reconnoistre quelque chose de superieur, qui est la force; celle-cy n'est pas des moindres. Monsieur Newton reconnoist l'equivalence des

Hypothese en cas des mouuements rectilineaires; mais a l'egard des Circulaires, il croit que l'effort que font les corps circulans de s'eloigner du centre ou de l'axe de la circulation fait connoistre leur mouuement absolu. Mais j'ay des raisons qui me font croire que rien ne rompt la loy generale de l'Equivalence. Il me semble cependant que vous meme, Monsieur, estiés autres fois du sentiment de M. Neuton à l'egard du mouuement circulaire.[50]

As this letter shows, Leibniz finds himself in a precarious situation, embracing the logical principle of kinematical relativity on the one hand, and the phenomenon of circular motion which demands the existence of absolute space, on the other. His "true motion," which differs from pure geometrical motion conceptually, is obviously an attempt at a compromise.

But Huygens is opposed to any compromise. Thus he writes in a letter dated August 24, 1694:

Pour ce qui est du mouvement absolu et relatif, j'ay admire vostre memoire, de ce que vous vous estes souvenu, qu'autrefois j'estois du sentiment de Mr. Newton, en ci qui regard le mouvement circulaire. Ce qui est vray, et il n'y a que 2 ou 3 que j'ay trouve celuy qui est plus veritable, duquel il semble que vous n'estes pas eloigne non plus maintenant, si non ence que vous voulez, que lorsque plusieurs corps ont entre eux du mouvement relatif, ils aient chacun un certain degre de mouvement veritable, ou de force, enquoy je ne suis point de vostre avis.[51]

Leibniz's reply of September 14, 1694, which brings this highly interesting exchange of ideas to an end, Huygens dying in 1695, shows his great interest in Huygens' solution of the problem of circular motion. He agrees that no special privilege attaches to circular motion as compared with uniform translational motion and that all reference systems should be treated as equivalent. In Leibniz's opinion it is merely the principle of simplicity that leads to the exclusive ascription of certain motions to certain bodies. No doubt this was borrowed by Leibniz from the realm of astronomy, where for many years it played an important role in the controversy between the Copernicans and their opponents. Leibniz not only realized the inherent similarity, or near identity,

[50] *Ibid.*, p. 639.
[51] Huygens, *Oeuvres complètes*, vol. 10, p. 609.

of the problem under discussion with the problem whether the Ptolemaic or the Copernican system is preferable, but he even composed a treatise, *Tentamen de motuum coelestium causis*,[52] whose intention is to show how the arguments with regard to the mechanical relativity of motion suggest the equivalence of the two rival cosmological systems. It seems that he originally intended to publish this work in Rome during his visit to the Holy City. But caution prevailed and he submitted only a *Promemoria*,[53] whose theoretical part begins with the statement: "Ut vero res intelligatur exactius, sciendum est Motum ita sumi, ut involvat aliquid respectivum et non posse dari phaenomena ex quibus absolute determinetur motus aut quies; constitit enim motus in mutatione situs seu loci."

We have mentioned Leibniz's last letter to Huygens which deals with the problem of absolute space. The letter is as follows:

. . . Comme je vous disois un jour à Paris qu'on avoit de la peine à connoistre le veritable sujet du Mouuement vous me répondîtes que cela se pouuoit par le moyen du mouuement circulaire, cela m'arresta; et je m'en souuins en lisant à peu près la même chose dans le liure de Mons. Newton; mais ce fut lorsque je croyois déja voir que le mouuement circulaire n'a point de privilege en cela. Et je voy que vous estes dans le meme sentiment. Je tiens donc que toutes les hypotheses sont equivalentes et lorsque j'assigne certains mouuements à certains corps, je n'en ay ny puis avoir d'autre raison, que la simplicité de l'Hypotheses croyant qu'on peut tenir la plus simple (tout consideré) pour la veritable. Ainsi n'en ayant point d'autre marque, je crois que la difference entre nous, n'est que dans la maniere de parler, que je tache d'accomoder a l'usage commun, autant que je puis, salva veritate. Je ne suis pas même fort elogne de la vostre, et dans un petit papier que je communiquay à Mr. Viviani, et qui me paroissoit propre à persuader Messieurs de Rome a permettre l'opinion de Copernic, je m'en accommodois. Cependant si vous estes dans ces sentimens sur la realité du mouuement, je m'imagine que vous deuriés en avoir sur la nature du corps de differens de ceux qu'on a coustume d'avoir. J'en ay d'assez singuliers et qui me paroissent demonstrés.[54]

What is this singular conception of the nature of bodies on the basis of which Leibniz here claims to have found the solu-

[52] Gerhardt, *Leibnizens mathematischen Schriften*, vol. 6, p. 144.
[53] *Ibid.*, p. 146.
[54] Huygens, *Oeuvres complètes*, vol. 10, p. 681.

tion of the problem of circular motion? We do not know. Leibniz does not explain his solution, either here or elsewhere as far as is known.

We are in a more fortunate position with regard to Huygens' solution of the same problem. How could Huygens maintain, in the light of certain dynamical effects such as the rise of centrifugal forces in circular motion, the kinematical principle of relative motion, and at the same time dispense with the existence of absolute space and motion?

In 1886 L. Lange drew attention to the possibility of finding Huygens' solution among his posthumous papers in the archives of Leyden. It was, however, only in 1920 that D. J. Korteweg and J. A. Schouten, having found in the Leyden archives four loose sheets written by Huygens, and all dealing with circular motion, published the solution. The fourth paper, in which Huygens summarized his solution, is quoted in part:

Diu putavi in circulari motu haberi veri motus "criterion" ex vi centrifuga. Etenim ad ceteras quidem apparentias idem fit sive orbis aut rota quaepiam me juxta adstante circumrotetur, sive stante orbe illo ego per ambitum ejus circumferar, sed si lapis ad circumferentiam ponatur projicietur circumeunte orbe, ex quo vere tunc et nulla ad aliud relatione eum moveri et circumgyrari judicari existimabam. Sed is effectus hoc tantummodo declarat impressione in circumferentiam facta partes rotae motu relativo ad se invicem in partes diversas impulsas fuisse, ut motus circularis sit relativus partium in partes contrarias concitatarum sed cohibitus propter vinculum aut connexum, an autem corpora duo inter se relative moveri possunt quorum eadem manet distantia?

Ita sane dum distantiae incrementum inhibetur, contrarius vero motus relativus per circumferentiam viget.

Plerique verum corporis motum statuunt cum ex loco certo ac fixo in spatio mundano transfertur, male nam cum infinite spatium undique extensum sit quae potest esse definitio aut immobilitas loci?

Stellas affixas, in Copernicano systemate, forsan revera quiescentes dicent. Sint sane inter se immotae sed omnes simul sumtae alterius corporis respectu quiescere dicentur, vel qua in re different a celerrime motis in partem aliquam ? nec quiescere igitur corpus nec moveri in infinito spatio dici potest, ideoque quies et motus tantum relativa sunt.[55]

[55] D. J. Korteweg and J. A. Schouten, *Jahresbericht der Deutschen Mathematiker-Vereinigung* 29, 136 (1920).

This may be translated:

For a long time I had thought that rotational motion by means of centrifugal forces contains a criterion for true motion. Indeed, with regard to other phenomena it is the same whether a circular disk or a wheel rotates near me, or whether I circle round the stationary disk. However, if a stone is put on the circumference this will be projected only if the disk rotates, and therefore I formerly thought that circular motion is not relative to any other body. Still, this phenomenon showed only that the parts of the wheel, owing to the pressure acting on the circumference, are driven in relative motion among themselves in different directions. Rotational motion is therefore only a relative motion of the parts, which are driven to different sides, but held together by a rope or other connection.

Now, is it possible to move two bodies relatively without changing their distance? This is indeed possible if an increase in their distance is prevented. An opposite relative motion exists on the circumference. Most people suppose that the true motion of a body consists in its being transferred from a certain fixed place in the universe. This is wrong; for if space is unlimited in all directions, what then is the definition of the immobility of a place? It will perhaps be said that the fixed stars in the system of Copernicus are really at rest; well, they may indeed be mutually immobile with respect to each other; but taken together, relative to what other body are they said to be at rest or in what respect are they to be distinguished from bodies moving very fast in a certain direction? It is therefore impossible to state that a body is at rest in infinite space, or that it moves therein; rest and motion are therefore only relative.

Thus Huygens thought he had discovered that the dynamical effect of the appearance of centrifugal forces is merely an indication of the relative motion of the different parts of the disk. Yet the relative motion of these parts can be transformed away by taking as a reference system just that system which has the same angular velocity (and the same origin) as the rotating disk. In this rotating coördinate system the parts of the disk are at rest. The dynamic effect, however, referred to this system, does not vanish: the "pressure" exerted by centrifugal forces has not been transformed away, as it should be were the centrifugal force but a dynamical effect of the relative motion of the particles. Huygens' explanation, therefore, certainly does not pass the test of modern scientific criticism. Nevertheless, it is a historical fact that Huygens, inspired by his sound scientific insight, was the first physicist who believed in the exclusive validity of a principle of kinematic as well as dynamic relativity, two hundred years before the rise of modern relativity.

CHAPTER 5

THE CONCEPT OF SPACE

IN MODERN SCIENCE

Nothing that Leibniz and Huygens had to say in criticism of Newton's concept of absolute space could prevent its acceptance. The letters that passed between Leibniz and Clarke, though widely read, were studied and discussed chiefly for their theological implications. With the gradual acceptance of the Newtonian system, and as the rival Cartesian theories fell out of grace, Newton's concept of absolute space became a fundamental prerequisite of physical investigation.

In this respect the words of John Keill, one of the early advocates of Newtonian physics in the University of Oxford, are

typical. In his second lecture, which he delivered at Oxford in
1700, he said:

> We conceive Space to be that, wherein all Bodies are placed, or,
> to speak with the Schools, have their Ubi; that it is altogether pene-
> trable, receiving all Bodies into itself, and refusing Ingress to nothing
> whatsoever; that it is immovably fixed, capable of no Action, Form
> or Quality; whose Parts it is impossible to separate from each other,
> by any Force however great; but the Space itself remaining immov-
> able, receives the Successions of things in motion, determines the
> Velocities of their Motions, and measures the Distances of the things
> themselves.[1]

Not only this sober, factual and scientific aspect of Newton's
conception of absolute space gained ground; the divinization of
space was acclaimed with no less enthusiasm by the early eighteenth century, as it conformed so very well with the general
outlook of the time for which science has become identical with
the study of the works of God. "Nature was rescued from Satan
and restored to God." [2] What wonder, therefore, that Joseph
Addison praised Newton's religious interpretation with the words:
"The noblest and most exalted way of considering this infinite
space is that of Sir Isaac Newton, who calls it the sensorium
of the Godhead." [3] The words of the Psalmist, "Coeli enarrant
gloriam Dei," could now be interpreted in a new sense.

> The spacious firmament on high
> With all the blue aethereal sky
> And spangled heavens, a shining frame,
> Their great Original proclaim.

The spread of such ideas was not confined to Europe. In the new
world Jonathan Edwards, metaphysician and divine, who still
one hundred years after his death was called "the greatest metaphysician America has yet produced," [4] expressed ideas about

---

[1] John Keill, *An introduction to natural philosophy* (London, 1745), p. 15.
[2] Basil Willey, *The eighteenth century background* (Chatto and Windus, London, 1949), p. 4.
[3] *The Spectator*, No. 565 (1714).
[4] Georges Lyon, *L'Idéalisme en Angleterre au XVIIIe siècle* (Paris, 1888), p. 406.

space which in their theological implications were most congenial to those expressed by Isaac Newton.

In England, in particular, Newton's inclusion of religious ideas in his system of physics was hailed as an outstanding achievement in natural philosophy. By analyzing the fundamental concepts of science, it was hoped, new material could be brought to light of avail in proving the existence of God. These proofs, based on the infinity and absoluteness of space, were to replace the traditional scholastic demonstrations which began to be considered as logically deficient. Such ideas were expressed by numerous authors during the first half of the eighteenth century, of whom we mention only Jacob Raphson,[5] John Jackson,[6] Joseph Clarke,[7] and Isaac Watts.[8]

The notion of absolute space triumphed on all fronts. Not only that, but, during the eighteenth century, attempts were made to demonstrate the logical necessity of the concept. Indeed, no less a man then Leonhard Euler grappled with the problem for more than thirty years. In his *Mechanica sive motus scientiae analytice exposita* Euler develops his mechanics on Newtonian lines and introduces the concept of absolute space and absolute motion in the spirit of the *Principia*. So his second definition reads: "Locus est pars spatii immensi seu infiniti in quo universus mundus consistit. Vocari hoc sensu acceptus locus solet absolutus, ut distinguatur a loco relativo, cuius mox fiet mentio."[9] To Euler, however, in his early work the question of the real existence of absolute space is a matter of indifference. Whether absolute space exists or not, it is only necessary to imagine such a space for the determination of absolute motion or absolute rest. But Euler

[5] Jacob Raphson, *De spatio reali seu ente infinito conamen mathematico-metaphysicum* (London, 1702).

[6] John Jackson, *The existence and unity of god, proved from his nature and attributes* (London, 1734).

[7] *Examination of Dr. Clarke's notion of space* (Cambridge, 1734).

[8] Isaac Watts, *Philosophical essays on various subjects* (London, ed. 2, 1736), "Essay I: A fair enquiry and debate concerning space whether it be something or nothing, God or a creature."

[9] L. Euler, *Mechanica sive motus scientiae analytice exposita* (St. Petersburg, 1736), p. 2.

changed his mind and in his "Réflexions sur l'espace et le temps" [10] emphasized the necessary existence of absolute space; for he had come to the conclusion that some real existing substratum is indispensable to the determination of motion. Since this substratum appears not to exist in the casual surrounding material, it must be space itself that exists in this capacity. "On en devroit plutôt conclure, que tant l'espace absolu, que le temps, tels que les Mathématiciens se les figurent, étoient des choses réelles, qui subsistent mêmes hors de notre imagination." [11] Euler's demonstration of the reality of absolute space on the basis of the law of inertia appears finally in his *Theoria motus corporum solidorum seu rigidorum*,[12] though the *Mechanica* already insists that the laws of motion presuppose the existence of absolute space. "Si hac significatione expositae voces accipiantur, vocari solent motus absolutus, quiesque absoluta. Atque hae sunt verae et genuinae istarum vocum definitiones, sunt enim accomodatae ad leges motus, quae in sequentibus explicabuntur." [13] Of course, Euler was not the first among Newton's successors who emphasized the intrinsic importance of the concept of absolute space for the formulation of the law of inertia. Maclaurin, in his *Account of Sir Isaac Newton's philosophical discoveries*, had stated explicitly: "This perseverence of a body in a state of rest or uniform motion, can only take place with relation to absolute space, and can only be intelligible by admitting it." [14] Euler did not confine himself to merely stating such an implication. If it were possible to demonstrate the logical necessity of the law of inertia, then, according to Euler, the logical necessity of absolute space would follow from it by implication. Thus, after he formulates the law of inertia, Euler attempts to give an a priori vindication of its necessity. The law is formulated in Axioma 2

[10] *Histoire de l'Académie Royale des Sciences et des Belles-lettres, 1748* (Berlin, 1750), vol. 4, p. 324.

[11] Reference 10.

[12] Rostock and Greifswald, 1765.

[13] Euler, *Mechanica*, p. 2.

[14] Colin Maclaurin, *Account of Sir Isaac Newton's philosophical discoveries* (London, 1748), book 2, chap. 1, sec. 9.

as follows: "Corpus, quod absolute quiescit, si nulli externae actioni fuerit subjectum, perpetuo in quiete perseverabit." [15] In the "Explicatio" immediately following we read, "cum enim in eo (i.e. elemento corporis) nulla insit ratio, cur in unam potius directionem moveri incipiat, quam in omnes alias, atque extrinsecus omnis causa motus adimatur, secundum nullam directionem motum concipere poterit. Nititur igitur quidem haec veritas principio sufficientis rationis."

It was a fairly common assumption of the time that by means of the principle of sufficient reason the law of inertia, and hence indirectly the existence of absolute space, could be demonstrated. Much the same reasoning occurs in D'Alembert's *Traité de dynamique* and also especially in Kant's *Metaphysical foundations of natural science*. But before this Kant had entertained a different opinion. In his youth Kant showed great interest in the natural sciences; in fact, from the time of his first work in 1747[16] the problem of space and motion occupied him constantly. In his *Principiorum primorum cognitionis metaphysicae nova dilucidatio* of the year 1755 Kant attempted to reconcile Newton with Leibniz. Agreeing with Leibniz's relational point of view, Kant sees in spatial relations not reflections of simply qualitative data given within the order of coexisting matter but rather mutual effects and interactions among bodies; since causal interdependence is not given with matter as such, but has been added and imparted by divine creation, space is an independent existent of absolute reality in the Newtonian sense. A similar compromise between Leibniz's metaphysics and Newton's physics is aimed at in the *Monadologia physica* of 1756. The metaphysician, according to Kant, asserts that all substantial reality is constituted of fundamental indivisible units or monads; the mathematician, on the other hand, claims that space is infinitely divisible; and the physicist finally applies mathematical space to metaphysical

---

[15] Euler, *Theoria motus*, p. 32.
[16] Kant, "Thoughts on the true estimation of living forces," in John Handyside, trans., *Kant's inaugural dissertation and early writings on space* (Chicago, 1929).

matter. This state of affairs is consistent only if space is not a substance but a phenomenon of relations among substances and if substance is but a center of action effecting other substances, and effected by them, through the mutual operation of forces. A simple substance "occupies" a larger or smaller space, not because it fills it up with a larger or smaller number of material parts, but because it exerts stronger or weaker forces of repulsion to prevent the approach of adjacent monads. Spatial magnitude is therefore only a measure of the intensity of acting forces exerted by the substance. In his *Neuer Lehrbegriff der Bewegung und Ruhe* Kant stresses again the pure relative character of space and writes:

Now I begin to see that I lack something in the expression of motion and rest. I should never say, a body is at rest, without adding with regard to what it is at rest, and never say that it moves without at the same time naming the objects with regard to which it changes its relation. If I wish to imagine also a mathematical space free from all creatures as a receptacle of bodies, this would still not help me. For by what should I distinguish the parts of the same and the different places, which are occupied by nothing corporeal? [17]

Yet five years later,[18] apparently under the influence of Euler,[19] Kant abandons this point of view and declares himself in favor of the Newtonian concepts of absolute space and absolute time. In his essay "On the first grounds of the Distinction of Regions in Space" Kant formulates his program as follows: "My aim in this treatise is to investigate whether there is not to be found in the judgments of extension, such as are contained in geometry, an evident proof that space has a reality of its own, independent of the existence of all matter, and indeed as the first ground of the possibility of the compositer ss of matter." [20] Kant thought

[17] Kant, *Gesammelte Werke* (Akademie Ausgabe; Berlin, 1905), vol. 2, p. 13.
[18] Kant, "Versuch den Begriff der negativen Grösse in die Weltweisheit einzuführen" (1763) in *Gesammelte Werke* (Akademie Ausgabe), vol. 2, p. 165.
[19] On Euler's influence on Kant, see H. E. Timerding, "Kant und Gauss," *Kantstudien* 28 (1923).
[20] Kant, "Von dem ersten Grunde des Unterschiedes der Gegenden im Raume" (1769) in *Gesammelte Werke* (Akademie Ausgabe), vol. 2, p. 375.

that here he had found an incontestable proof of the existence and reality of absolute space, independent of the existence of matter. As the first ground for the possibility of the composite-ness or disposition of matter, space is endowed with a reality of its own. Kant bases his proof on the distinction between left and right. It is observed, he says, that the intrinsic relations among the individual parts of our left hand with regard to each other are the same as in our right hand; and yet evidently a fundamental distinction makes it impossible to substitute one hand for the other. Now, if this fundamental difference cannot be explained as being merely the appearance of different relation in the order or disposition of the parts with respect to each other, it can be accounted for only by the assumption of a different disposition with regard to absolute space. Thus, absolute space must be introduced as a fundamental metaphysical notion necessary for the explanation of this phenomenon.

H. Weyl, who shows that mathematically the root of this distinction is of a purely combinatorial character (a permutation of given linearly independent vectors determines the "sense" of rotation, as for example in left- or right-handed coördinated sys-tems), says of Kant's argumentation:

Kant finds the clue to the riddle of left and right in transcendental idealism. The mathematician sees behind it the combinatorial fact of the distinction of even and odd permutations. The clash between the philosopher's and the mathematician's quest for the roots of the phenomena which the world presents to us can hardly be illustrated more strikingly.[21]

The left side of a straight line can be interchanged with its right side by rotating the line in a plane, the clockwise direction on a surface can be interchanged with an anticlockwise direc-tion by moving the surface in three-dimensional space (turning over), a left-hand screw with a right-hand screw — or the left hand with the right hand — by "moving" the object in four-dimensional space. It is clear, therefore, that mathematically no

[21] H. Weyl, *Philosophy of mathematics and natural science* (Princeton University Press, Princeton, 1949), p. 84.

essential mark distinguishes one sense from the other. In fact, until
1956, when for the first time the conservation of parity was called
into question, a point to be discussed later on, it was generally
believed that all laws of nature are invariant with respect to an
interchange of right and left. True, certain asymmetries in chemical or biological phenomena had been recognized. Weyl, for example, mentioned the following fact:

That homo sapiens contains a screw turning the same way in all
individuals is proved in a rather horrid fashion by the fact that man
contracts a metabolic disease called phenylketonuria leading to amentia
when a certain quantity of levo-phenylalanine is added to his food,
whilst the dextro form has no such disastrous effect.[22]

It seemed, however, that phenomena of this kind have no deeper
significance and nobody would have used them as a proof for the
existence of absolute space.

But back to Kant. It is in his view only immediate intuition
that distinguishes between left and right, a difference that cannot
be formulated conceptually. Furthermore, it is immediate intuition that forms our general conceptions in geometry and makes
their statements evident. In this intuition rests the proof for the
reality of absolute space. "In den anschauenden Urteilen, derglei-
chen die Messkunst enthaelt, ist der Beweis zu finden, dass der
absolute Raum unabhängig von dem Dasein aller Materie und
selbst als der erste Grund ihrer Zusammensetzung eine eigene
Realität habe." [23] The idea that intuition lies at the basis of our
geometric cognition brings about a radical change in Kant's attitude toward these questions. The problem of space now appears
to Kant in a new light. It ceases to be a problem of physics and
becomes an integral part of transcendental philosophy. To Kant
from now on space is a condition of the very possibility of experience. In the inaugural dissertation "De mundi sensibilis atque
intelligibilis forma et principiis," [24] the concepts of absolute space
and absolute time are considered to be merely conceptual fic-

[22] *Ibid.*, p. 208; see also H. Weyl, *Symmetry* (Princeton University Press, Princeton, 1952), pp. 16–38.
[23] Kant, *Gesammelte Werke* (Akademie Ausgabe), vol. 16.
[24] Kant, "Dissertation on the form and principles of the sensible and intelligible world" (1770), in Handyside, reference 16, p. 33.

tions, a mental scheme of constructed relations of coexistence and sequence among sense particulars. Not itself arising out of sensations, the concept of space is a pure intuition, neither objective nor real, but subjective and ideal.

Kant's critical theory of space, like his philosophy in general, was greatly influenced by the English empiricists Locke, Berkeley, and Hume, and their analytical investigations into the formation of ideas. For Locke, extension, figure, size, and motion, in contrast to color, sound, and taste, were primary qualities, inherent in the object and independent of the perceiving subject. Berkeley, in an attempt to explain our conception of space, in his *New theory of vision*[25] reduced visual space to visual signs of tangible space, arguing that visual ideas are within the mind, whereas tangible space, in his view, needs no explanation. Distances, sizes, or figures, in consequence, are not "seen" or perceived, but inferred by the mind, experience having shown that certain visual sensations are related to certain tactile relations. In his other works, however, Berkeley asserts that tangible ideas (objects) are also within the mind, but he does not reëxamine his exposition of the perception of visual space in the light of his new position. Notwithstanding this inconsequence, extension, size, and figure appear in Berkeley's philosophy as secondary qualities.

In his treatise *Concerning the principles of human knowledge* Berkeley describes how according to his empiristic view the concept of space is formed by the perception of extension, the notion of space being but an abstract idea of extension. For like other general ideas it is formed in the human mind by abstraction from sense perceptions relating to bodies. Newton's notion of absolute space, which contains all bodies and is retained if all bodies are thought away, is in Berkeley's view a false hypostatization of an abstraction. Only particular spaces, corresponding to extensions perceived by our senses through their colors, figures, and tactile

[25] Berkeley, *A new theory of vision and other writings* (Dent, London, 1938), p. 37.

qualities, are to be admitted. The notion of empty space, however, is a mere verbal expression of a state of empirical facts.

When I excite a motion in some part of my body, if it be free or
without resistance, I say there is space: but if I find a resistance, then
I say there is body: and in proportion as the resistance to motion is
lesser or greater, I say the space is more or less pure. So that when I
speak of pure or empty space, it is not to be supposed, that the word
space stands for an idea distinct from, or conceivable without body
and motion. Though indeed we are apt to think every noun substan-
tive stands for a distinct idea, that may be separated from all others:
which hath occasioned infinite mistakes.[26]

The adoption of such an extreme subjective idealism is the
final conclusion of an empiristic approach, according to which all
our knowledge, furnished by experience, goes back to elementary
sense data subjected to the reflection of the mind.

Kant's position in this context is best characterized by his own
words: "Although all knowledge begins with experience, it does
not necessarily all spring from experience." The object of sensation is not identical with the object of thought. This applies in
particular to the conception of space, which, according to Kant,
is a form of intuition, instrumental in the process of cognition as
an ideal organizer of the contents of sensations.

Kant's claim of the transcendental ideality of space (and
time) is expounded both in his *Prolegomena* and in the *Critique
of pure reason,* where its discussion, under the name of Transcendental Aesthetic, plays a fundamental role in his epistemology.
It is well known how Kant tries to demonstrate that the immediate object of perception is rooted partly in external things and
partly in the apparatus of our own perception. The first component, due to the "thing-in-itself," is called the "sensation," and
the second component, the "form" of the phenomenon. It is this
second component that brings order into the amorphous manifold
of our sensations; it is an a priori element of our perception,
antecedent to all experience. It is universal, since it does not

[26] Berkeley, *Principles of human knowledge* in *A new theory of vision,*
p. 173.

depend on the particular data of our sensation. As a pure form of sensibility Kant calls this component "pure intuition" (*reine Anschauung*). There are two pure intuitions: space and time.

Four metaphysical arguments based on the nature of space and time, and one transcendental argument derived from the special character of Euclidean geometry, are set forth to prove the contention.

The metaphysical exposition begins with the classical words: "Space is not an empirical notion which has been derived from external experience," [27] because, according to Kant, any possible sensation referred to something external presupposes the perception of space. The second argument stresses the fact that space is a necessary perception a priori, underlying all external perception, because we cannot imagine that there is no space, although we can well imagine that there are no objects found in space. Thirdly, space is no discursive or general notion (as "animal" or "table"), for, speaking of spaces, we mean merely parts of one and the same space. Finally, space is conceived as an infinite magnitude. "Now a notion must be conceived, indeed, as common to an infinite number of different possible individuals," that is to say, the notion has to be contained in an infinite number of particulars subsumed by it. This relation does not hold in the case of the concept of space. Consequently, space is an intuition, a priori, and not a notion.

The transcendental argument is based on Kant's characterization of geometry as synthetic and yet a priori; geometric propositions are, in his view, apodictic, that is, they bring with them their own necessity, but they are not judgments of experience. Geometry can be known a priori without being a mere tautology, only because it lies at the basis of our perception.

Kant's metaphysical exposition tries to demonstrate that space and time are conditions under which sense perception operates. The a priori ideas of space and time are not images which cor-

<hr />

[27] Kant, *The critique of pure reason* (trans. by J. H. Stirling; Edinburgh, 1881), p. 141.

respond to external objects. In fact, there is no object in the external world called space. It is not an object of perception, it is a mode of perceiving objects. Nevertheless, Kant explicitly opposes the relational theory of space and endows the form of perception with existence independent of the particular bodies contained in it. In the *Critique of pure reason* he says:

> Space does not represent any property of things in themselves, nor does it represent them in their relation to one another. That is to say, space does not represent any determination that attaches to the objects themselves, and which remains even when abstraction has been made of all the subjective conditions of intuition.[28]

It is easy to understand that Kant's doctrine of the transcendental ideality of space, hailed in its days as one of the greatest achievements in contemporary philosophy, was greatly influential on the course of idealistic philosophy and had its repercussions also on psychology. Indeed, Lotze's theory of "local signs," Brentano's psychological researches, and Stumpf's investigation into the origin of space perception show clearly how Kant's epistemological inquiries were bound to be diverted into the field of the psychology of senses.

Since a discussion of the psychological origin of our space conception is outside the scope of our topic, we need not enter here into the details of A. Bain's, J. Mueller's, or Helmholtz's important contributions toward a clarification of this complicated problem, nor are we here concerned with the history of the nativistic or of the empiristic theories on the formation of space perception. It is, however, important for us to note that Helmholtz, one of the chief proponents of the empiristic school, has shown that the two parts of Kant's doctrine, the metaphysical and the transcendental expositions, are not so closely connected as was originally assumed. Although Helmholtz[29] does not reject

[28] N. K. Smith, trans., *Immanuel Kant's Critique of pure reason* (Macmillan, London, 1950), p. 71.

[29] H. von Helmholtz, "Ueber die Tatsachen welche der Geometrie zugrunde liegen," *Göttinger gelehrte Nachrichten* (1868), pp. 193–221. See also his "Ueber den Ursprung und die Bedeutung der geometrischen Axiome," *Vorträge und Reden* (Brunswick, ed. 5, 1903), vol. 2, pp. 1–31.

in principle the metaphysical argumentation, he strongly objects to the assumption of the a priori nature of Euclidean geometry. Space, as a pure form of intuition, leads, according to Helmholtz, to one single conclusion: that all objects of the external world must necessarily be endowed with spatial extension. The geometric character of this extension, however, is in his view purely a matter of experience. Helmholtz's opinion, imbued with the recognition of the validity of non-Euclidean geometry, may, on the whole, still be taken today as representative of the attitude of physicists toward Kant's doctrine of space.[30] Modern psychology denies the allegation that a determinate metric space must be assumed to exist as an organizing element of cognition. It should also be recalled that modern logic no longer respects Kant's double dichotomy of judgments into a priori and a posteriori, analytic and synthetic, as a fundamental classification for epistemological research. It is certainly a very conservative statement to say that the line between a priori and a posteriori has been drawn at different places at different times.

It is interesting to note how little the actual progress of the science of mechanics was affected by general considerations concerning the nature of absolute space. Among the great French writers on mechanics, Lagrange, Laplace, and Poisson, none of them was much interested in the problem of absolute space. They all accepted the idea as a working hypothesis without worrying about its theoretical justification. In reading the introductions to their works, one discovers that they felt that science could very well dispense with general considerations about absolute space. It is interesting to note that the *Encyclopédie* of Diderot and d'Alembert expresses much the same view. In the fifth volume, under the heading *"Espace,"* we read:

Cet article est tiré des papiers de M. Formey, qui l'a composé en partie sur le recueil des Lettre de Clarke, Leibnitz, Newton, Amsters. 1740. & sur les inst. de Physique de madame du Châtelet. Nous ne prendrons point de parti sur la question de l'espace; on peut voir,

[30] Cf. Viktor Henry, "Das erkenntnistheoretische Raumproblem," *Kantstudien*, Ergänzungsheft No. 34 (1915).

partout ce qui a été dit au mot Elémens des Sciences, combien cette question obscure est inutile à la Géométrie & à la Physique.[31]

It may even be claimed that this absence, so far from being a hindrance to mechanics in the eighteenth and early nineteenth century, in a certain degree facilitated the development of this science. In England, by the middle of the nineteenth century, it became clear that the concept of absolute space was useless in physical practice. In that country the great success of Newtonian physics led to the paradoxical situation of the adherence to the concepts of absolute time and absolute space, on the one hand, and their absence from practical physics, on the other. Clerk Maxwell's remarks on absolute space in his *Matter and motion* are characteristic:

Absolute space is conceived as remaining always similar to itself and immovable. The arrangement of the parts of space can no more be altered than the order of the portions of time. To conceive them to move from their places is to conceive a place to move from itself. But as there is nothing to distinguish one portion of time from another except the different events which occur in them, so there is nothing to distinguish one part of space from another except its relation to the place of material bodies. We cannot describe the time of an event except by reference to some other event, or the place of a body except by reference to some other body. All our knowledge, both of time and space, is essentially relative.[32]

In 1885 an important attempt to find a way out of the paradoxical situation (that is, the adherence to the concept of absolute space on the one hand and its absence from practical physics on the other) was made by Ludwig Lange.[33] Lange thought that he had found the way to eliminate the concept of absolute space from the conceptual foundation of physics. In his view, the essential (today he would say the operational) content of the law of inertia, and with it of the whole of mechanics, retains its full

[31] Diderot and d'Alembert, *Encyclopédie, ou dictionnaire raisonné des sciences, des arts et des métiers*, vol. 5 (1755), p. 949.
[32] J. C. Maxwell, *Matter and motion*, reprinted, with notes and appendices by Sir Joseph Larmor (Dover, New York, n.d.), p. 12.
[33] L. Lange, "Ueber die wissenschaftliche Fassung des Galileischen Beharrungsgesetzes," *Ber. kgl. Ges. Wiss., Math.-phys. Kl.* (1885), pp. 333–351.

physical meaning if the somewhat "ghostly" idea of an absolute space is replaced by the concept of an "inertial system." Let there be given a mass-point A whose motion is arbitrary (even curvilinear). Then it is always possible to move a coördinate system S in such a manner that A moves relative to S along a straight line a. If in addition a second point B and a third point C are given with arbitrary motions, the coördinate system can still be moved in such a way that all three mass-points, relative to S, move along straight lines a, b, c. Now, three is the maximum number of mass-points for which in general the construction of a coördinate system S is possible in which the points move along straight lines. If we now assume that the three points A, B, C, being projected from the same origin, are left to themselves (that is, are not subjected to any forces), the corresponding coördinate system S, relative to which the three points describe three different straight lines, is defined as an "inertial system." The physical contents of the law of inertia, according to Lange, is equivalent to the contention that any fourth mass-point, left to itself, also moves along a straight line relative to S. In short, an "inertial system" is a coördinate system in respect to which Newton's law of inertia holds. Lange's suggestion of eliminating the idea of absolute space by introducing the concept of an "inertial system" was hailed by his contemporaries as an outstanding contribution to the foundations of physics. Seeliger[34] thought that it was possible to compare Lange's inertial system with the empirical coördinate system used in astronomy and stated that the relative motion between these two systems is less than 2 seconds of arc within the span of a century.

Toward the end of the nineteenth century it became obvious that absolute space evaded all means of experimental detection. Mach showed that the assumption of absolute space for the explanation of centrifugal forces in rotational motion was unnecessary. In the *Science of mechanics* he writes:

[34] H. Seeliger, "Ueber die sogenannte absolute Bewegung," *Sitzber. Münchener Akad. Wiss.* (1906), p. 85.

Newton's experiment with the rotating vessel of water simply informs us, that the relative rotation of the water with respect to the sides of the vessel produces no noticeable centrifugal forces, but that such forces are produced by its relative rotation with respect to the mass of the earth and the other celestial bodies. No one is competent to say how the experiment would turn out if the sides of the vessel increased in thickness and mass till they were ultimately several leagues thick.[35]

Mach's modification of the traditional interpretation of Newton's pail experiment and his objection to accepting the experiment as a proof of the existence of absolute space are the result of his conviction that all metaphysical concepts have to be eliminated from science. In the foreword to the first edition of his *Die Mechanik in ihrer Entwicklung*[36] he writes: "Vorliegende Schrift ist kein Lehrbuch zur Einübung der Sätze der Mechanik. Ihre Tendenz ist vielmehr eine aufklärende oder, um es noch deutlicher zu sagen, eine antimetaphysische." The very idea of an absolute space, that is, of an agent that acts itself but cannot be acted upon, is, in his view, contrary to scientific reasoning. Space as an active cause, both for translational inertia in rectilinear motion and for centrifugal forces in rotational motion, has to be eliminated from the system of mechanics.

With reference to Newton's theory of space, Mach is ready to accept only the idea of relative spaces (see p. 98), which approximate inertial systems. In the concluding passage of his foreword to the seventh edition (1912) of his *Mechanik* Mach says: "Bezüglich der Begriffsungetüme des absoluten Raumes und der absoluten Zeit konnte ich nichts zurücknehmen. Ich habe hier nur deutlicher als vorher gezeigt, dass Newton zwar manches über diese Dinge redet, aber durchaus keine ernste Anwendung von denselben gemacht hat. Sein Coroll. V. (*Principia*, 1687, p. 19) enthält das einzig praktisch brauchbare (wahrscheinlich angenährte) Inertialsystem."[37]

[35] E. Mach, *The science of mechanics* (trans. by T. J. McCormack; Chicago, 1902), p. 232.
[36] E. Mach, *Die Mechanik in ihrer Entwicklung* (Leipzig, 1883).
[37] *Ibid.* (Leipzig, ed. 8, corresponding with ed. 7, 1921), p. x.

The elimination of what Mach calls "the conceptual monstrosity of absolute space" (*das Begriffsungetüm des absoluten Raumes*) is achieved, in his view, by relating the unaccelerated motion of a mass-particle not to space as such, but to the center of all masses in the universe. The assumption of an intrinsic functional dependence between inertia and a large-scale distribution of matter closes for him the series of mechanical interactions without resorting to a metaphysical agent. "Ueber den absoluten Raum und die absolute Bewegung kann niemand etwas aussagen, sie sind blosse Gedankendinge, die in der Erfahrung nicht aufgezeigt werden können." The very fact that absolute space and absolute motion are physically imperceptible, even if their objective existence may be admitted philosophically, characterizes them — in Mach's terminology — as "metaphysical" and demands their elimination from exact science. In the fourth edition of *Die Mechanik* Mach summarizes his ideas concerning space in a very clear statement (which curiously is omitted in the later editions):

Für mich gibt es überhaupt nur eine relative Bewegung und ich kann darin einen Unterschied zwischen Rotation und Translation nicht machen. Dreht sich ein Körper relativ gegen den Fixsternhimmel, so treten Fliehkräfte auf, dreht er sich relativ gegen einen anderen Körper, nicht aber gegen den Fixsternhimmel, so fehlen die Fliehkräfte. Ich habe nichts dagegen, wenn man die erstere Rotation eine absolute nennt, wenn man nur nicht vergisst, dass dies nichts anderes heisst, als eine relative Drehung gegen den Fixsternhimmel. Können wir vielleicht das Wasserglas Newtons festhalten, den Fixsternhimmel dagegen rotieren, und das Fehlen der Fliehkräfte nun nachweisen? Der Versuch ist nicht ausführbar, der Gedanke überhaupt sinnlos, da beide Fälle sinnlich voneinander nicht zu unterscheiden sind. Ich halte demnach beide Fälle für denselben Fall und die Newtonsche Unterscheidung für eine Illusion.

It is obvious that these words may be considered as the earliest proclamation of the principle of general relativity and, in fact, they have been interpreted as such.[38]

Mechanics, it seemed, had to give up the notion of absolute space. Under the stress of these circumstances it was suggested

[38] For example, by W. Wien, *Die Relativitätstheorie* (Leipzig, 1921), p. 31.

by Drude and Abraham, to mention only these names, that the ether, the carrier of electromagnetic waves, should be identified with absolute space. If the ether as an absolute reference system could be demonstrated, the notion of absolute space could be saved. Indeed, one of the most important experiments to this end, the Michelson-Morley experiment, was in 1904 interpreted by Lorentz in this sense. His interpretation fulfilled all physical requirements. As is well known, according to Lorentz every body moving with reference to the motionless ether or absolute space undergoes a certain contraction in the dimension parallel to the motion. However, the Michelson-Morley experiment served as the starting point for the development of the theory of relativity and was interpreted by Einstein on entirely different lines, adverse to the acceptance of absolute space. It was understood that both interpretations give a complete explanation of all observations known at the beginning of the twentieth century. An *experimentum crucis* could not decide between these two theories. As Laue explains the situation in 1911:

A really experimental decision between the theory of Lorentz and the theory of relativity is indeed not to be gained, and that the former, in spite of this, has receded into the background, is chiefly due to the fact, that close as it comes to the theory of relativity, it still lacks the great simple universal principle, the possession of which lends the theory of relativity . . . an imposing appearance.[39]

Epistemologically, Lorentz's theory shows its unsatisfactory character by the fact that it ascribes to the ether or absolute space certain definite effects which by their very assumed existence preclude any possible observation of the ether. Similarly, all other experiments to identify the ether as a privileged system of reference had to go by the board. Physics, and not only mechanics, was ready to abandon the concept of absolute space altogether. Poincaré's words, "Whoever speaks of absolute space uses a word devoid of meaning," [40] became an accepted truth.

[39] Quoted from E. Cassirer, *Einstein's theory of relativity considered from the epistemological standpoint* (Chicago, 1923), p. 376.
[40] H. Poincaré, *Science and method* (trans. by F. Maitland; London, 1914), p. 93.

Yet space, no longer an absolute entity, retained one property in common with such an entity: it was Euclidean in nature. Even in the theory of special relativity, the space-time continuum by which every observer identifies the events in his physical world was held to be Euclidean, or pseudo-Euclidean, if the Minkowski representation is adopted. The question whether the space of experience was Euclidean or not was already a subject of discussion before the rise of general relativity. To Newton and his immediate successors, with no alternative before them, absolute space was naturally thought to be Euclidean. The discovery of non-Euclidean geometry led to the elimination of this last traditional characteristic of space, and modern physics came finally to base its conception of space upon the Riemann notion of an n-dimensional manifold.

It is an exciting story, and begins with Euclid and his fifth postulate, later simply called the "parallel axiom," which reads as follows:

That, if a straight line falling on two straight lines make the interior angles on the same side less than two right angles, the two straight lines, if produced indefinitely, meet on that side on which are the angles less than the two right angles.[41]

Or, in an equivalent formulation, it states that in a given plane through a given point not more than one parallel to a given line exists.[42] That this postulate is not needed for the demonstration of the first 28 theorems of the *Elements* was noted early. In antiquity it was thought possible to prove the postulate on the basis of the other postulates. From Ptolemy and Proclus to Nasiraddin-at-Tusi, the Persian editor of the *Elements,* and John Wallis, down to Lambert and Legendre, all attempts at such a proof failed. Of all the agelong attempts to solve the problem, the most remarkable is that of Girolomo Saccheri. In his *Euclides ab omne naevo vindicatus*[43] Saccheri tries to show that a con-

[41] Euclid, *The elements* (trans. by Sir Thomas Heath; St. John's College Press, Annapolis, 1947), vol. 1, p. 202.
[42] The existence of at least one parallel can be proved by the other postulates.
[43] Milan, 1733.

tradiction results if the postulate in question is replaced by another. Today we know that no contradiction with the other postulates would have arisen if Saccheri had not used unawares an assumption that is in fact equivalent to the fifth postulate. His book had a great influence on subsequent investigations into the nature of the postulate. The problem attracted many first-class mathematicians. One of the greatest of them, Carl Friedrich Gauss, seems to have recognized the logical possibility of a non-Euclidean geometry even before Lobachevski and Bolyai came out with their sensational discoveries. By the end of the first half of the nineteenth century it was clear that the fifth postulate could not be deduced from the others, since its negation led to no contradictions with the other postulates. Further, Klein succeeded in showing that by using a Euclidean model of non-Euclidean geometry, that is, by systematically interpreting non-Euclidean geometric terms by Euclidean terms, non-Euclidean geometry is certainly as consistent as Euclidean geometry. So Euclidean geometry stood as one system among others with no privileged position, at least from the point of view of logic.

Our interest here is not in non-Euclidean geometry as such, but in the remarkable effect it had on the concept of space in modern physics. Not only did it lead to a fuller understanding of the hypothetical nature of pure axiomatic geometry, and so to an understanding of the nature of mathematics in general, but it led also — and this is no less important — to the clarification of the concept of physical space as opposed to the concept of mathematical space. With the discovery of non-Euclidean geometry it became clear that there were no a priori means of deciding from the logical and mathematical side which type of geometry does in fact represent the spatial relations among physical bodies. It was natural, therefore, to appeal to experiment and to find out whether the question of the true geometry could be settled a posteriori.

Once the validity of non-Euclidean geometry was recognized, the question arose whether the space of physics was Euclidean or not. In the vanguard of the attack on this problem was F. K. Schweikart, professor of law at the University of Marburg, who, according to the historians of non-Euclidean geometry, must be reckoned among the first independent discoverers of this science. Schweikart published his geometric system under the title *Astralgeometrie,* intending to indicate thereby that only by experiments or observations on an astronomical scale could the difference between Euclidean geometry and his own geometry be detected. Better known is Gauss's attempt to find out whether the space of experience is Euclidean or not. He tried to measure directly by an ordinary triangulation with surveying equipment whether the sum of the angles of a large triangle amounts to two right angles or not. Accordingly, he surveyed a triangle formed by three mountains, the Brocken, the Hoher Hagen, and the Inselberg with sides measuring 69, 85, and 107 km. Needless to say, he did not detect any deviation from 180° within the margin of error and thus concluded that the structure of actual space is Euclidean as far as experience can show.

This was the first accurate survey of a geodetic triangle on a very large scale, and it required no little work. Its outcome must have been somehow disappointing for Gauss. The negative result, the fact that no deviation from Euclidean geometry has been ascertained, was not conclusive, that is, it could not serve either to prove or to disprove decisively his ideas about space, which had been in his mind already for a couple of years. For as early as 1817 he wrote to H. W. M. Olbers:

> I become more and more convinced that the necessity of our geometry cannot be demonstrated, at least neither by, nor for, the human intellect. In some future life, perhaps, we may have other ideas about the nature of space which, at present, are inaccessible to us. Geometry, therefore, has to be ranked until such time not with arithmetic, which is of a purely aprioristic nature, but with mechanics.[44]

Gauss's experimental investigation of the geometric structure of space was based, as we see, on his conviction, suggested

[44] K. F. Gauss, *Werke* (Königliche Gesellschaft der Wissenschaften zu Göttingen; Leipzig, 1863–1903), vol. 8, p. 177.

by the acknowledged validity of non-Euclidean geometry, that geometry is essentially different from arithmetic and analysis. Whereas the last two branches of mathematics are based on the idea of pure number, and remain, therefore, purely rational knowledge, geometry becomes an empirical science inasmuch as it requires experimental investigation. In a letter to Bessel [45] Gauss wrote that we have to admit that number is a product of the mind but space has a reality outside the mind whose laws we cannot prescribe a priori. Gauss seemed to have realized that his conception of space has far-reaching epistemological consequences. It was perhaps the anticipation of an impending conflict with orthodox philosophy that made him guard his secret most carefully for a number of years, fearing "the clamor and cry of the blockheads." Only in 1844 he wrote to his friend Schumacher, the director of the observatory in Kiel and editor of the *Astronomische Nachrichten,* the pungent remarks:

> You see the same sort of things [mathematical incompetence] in the contemporary philosophers Schelling, Hegel, Nees von Essenbeck, and their followers; don't they make your hair stand on end with their definitions? Read in the history of ancient philosophy what the big men of that day — Plato and others (I except Aristotle) — gave in the way of explanations. But even with Kant himself it is often not much better; in my opinion his distinction between analytic and synthetic propositions is one of those things that either run out in a triviality or are false.[46]

To Gauss, non-Euclidean geometry, or, as he called it, "anti-Euclidean" geometry, was logically impeccable but experiment seemed to preclude its application to physical space. Lobachevski, who independently shared this view, wrote in his *Neue Anfangsgründe der Geometrie:*

> The futility of all these efforts of the last two thousand years since the time of Euclid made me suspect that in geometry the concepts themselves do not imply the truth whose proof we sought and whose

---

[45] *Ibid.,* p. 201.
[46] Quoted from E. T. Bell, *Men of mathematics* (Simon and Schuster, New York, 1937), p. 240.

vindication, like the vindication of other natural laws, can be achieved only through experience, as for example by astronomical observation.[47]

If physical space were different from Euclidean space, he thought, the difference could be established only by means of large-scale observations.

For the benefit of the reader who has a rudimentary knowledge of non-Euclidean geometry, an astronomical method for determining the *space constant k*, on the assumption of hyperbolic geometry, will be explained. The numerical value of k depends, of course, on the arbitrary unit of length employed, but k may be used itself as a natural unit of length.

Let A and B (Fig. 3) be two opposite positions of the earth on its annual orbit around the sun S. Let F be a star, whose parallax

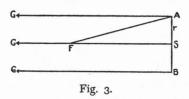

Fig. 3.

is defined to be the angle AFS subtended by the radius r of the earth's orbit. For a direct measurement of the parallax AFS the transit circle can be employed. On the assumption of Euclidean geometry the parallax is then given by the angle $\frac{1}{2}\pi - SAF$. A second method of measuring the parallax consists in comparing the position of F with that of another star G whose distance is taken to be very much greater than that of F. By measuring the angle GAF, the parallax, on the basis of Euclidean geometry, is found immediately, since it is equal to GAF. However, if space is assumed to be of hyperbolic structure, the two methods explained will yield different results, since the sum of the two angles GAF and FAS is different from $\frac{1}{2}\pi$. In fact, this sum is just the *angle of parallelism* $\Pi(r)$, corresponding to the radius r of

[47] N. I. Lobachevski, "Neue Anfangsgründe der Geometrie mit einer vollständigen Theorie der Parallellinien," *Kasaner Gelehrten Schriften* (1835–1838), p. 67.

the earth's orbit. Let δ be defined by the equation $2\delta = \frac{1}{2}\pi - \Pi(r)$. Then according to a fundamental theorem of hyperbolic geometry we have

$$e^{-r/k} = \tan \tfrac{1}{2}\Pi(r) = \tan \tfrac{1}{2}(\tfrac{1}{2}\pi - 2\delta) = \tan (\tfrac{1}{4}\pi - \delta) = \frac{1 - \tan \delta}{1 + \tan \delta}.$$

Taking natural logarithms we obtain

$$\frac{r}{k} = \ln \frac{1 + \tan \delta}{1 - \tan \delta},$$

and finally that r/k is approximately equal to 2 tan δ. A calculation of δ, on the basis of the two different methods of parallax measurement, in combination with the knowledge of the radius r of the earth's orbit, determines the space constant of hyperbolic space. A finite (real) value of k could be interpreted in favor of the hypothesis of hyperbolic space structure.

Lobachevski used a triangle whose base was the diameter of the earth's orbit and whose apex was the star Sirius, whose parallax he assumed to be 1.24″ according to a determination made by Comte d'Assa-Montdardier.[48] Lobachevski's data were wrong and it was only in 1838 that Bessel succeeded in measuring the first parallax of a star (61 Cygni; 0.45″). The true value of the parallax of Sirius is less than 0.40″, in other words, less than a third of the value accepted by Lobachevski.

Like Gauss's, Lobachevski's attempt to prove the non-Euclidean structure of space empirically came to nothing. So he concluded that Euclidean geometry alone was of importance for all practical purposes. Thus he writes:

However it be, the new geometry whose foundations are laid in this work, though without application to nature, can nevertheless be the object of our imagination; though not used in real measurements, it opens a new field for the application of geometry to analysis and vice versa.[49]

[48] D'Assa-Montdardier, *Mémoire sur la détermination de la parallaxe des étoiles* (Paris, 1828).
[49] Lobachevski, "Neue Anfangsgründe der Geometrie," p. 24.

The problem of the relevance of non-Euclidean geometry to physical space became a subject of controversy among scientists and philosophers of science, especially after the posthumous publication of Riemann's great treatise "On the hypotheses which lie at the basis of geometry." [50] In this treatise Riemann presented an analytic approach to non-Euclidean geometry, in contrast to the axiomatic approach of his predecessors. Analyzing the mathematical properties of a manifold of undefined objects, called points, which are determined by a set of coördinates, Riemann stressed for the first time in the history of mathematics the important distinction between the unlimited and the infinite. Since the time of Gassendi space, as a homogeneous continuum, was always thought of as unlimited, a boundary being obviously a singularity, mathematically speaking. Riemann showed that homogeneity and finiteness are compatible. His generalization of Gauss's theory of surfaces, culminating in the concept of "curved space," made it clear that the space of Euclidean geometry and the space of the geometry of Lobachevski and Bolyai were only special cases of the generalized space, that is, spaces of constant zero curvature or constant negative curvature. By introducing an appropriate metric Riemann was able to show also that a space of constant positive curvature, a so-called "spherical" space,[51] is conceivable.

This mathematical analysis of the structure of space, initiated by Gauss and Riemann, is of such paramount importance for the formation of modern space conceptions, both in mathematics and in physics, that a more detailed account of these investigations has to be given. Riemann's paper "On the hypotheses which lie at the basis of geometry," written at the age of only 28 years,

[50] G. F. B. Riemann, "Ueber die Hypothesen, welche der Geometrie zu Grunde liegen" (Habilitationsschrift, 1854), *Abhandl. kgl. Ges. Wiss. zu Göttingen* 13 (1868); see also H. Weber, ed., *Collected works of Bernhard Riemann* (Gesammelte mathematische Werke), (Dover, New York, ed. 2, 1953), p. 272.

[51] The surface of a sphere is a two-dimensional "space" of constant positive curvature.

became the foundation of a general theory of space. Furthermore, it gave a new impetus to the important development of modern tensor analysis, which, originally confined in its applications to the treatment of problems in elasticity, became through the works of Ricci, Beltrami, Christoffel, Lipschitz, Bianchi, Weyl, and Einstein an indispensable tool for higher mathematics as well as for theoretical physics.

Riemann successfully generalized Gauss's theory of the intrinsic geometry on a surface. Gauss's interest in geodesy, cartography, and allied branches of applied mathematics drew his attention to the problem of how far geometric properties on curved surfaces can be expressed without resorting to the geometry of the embedding three-dimensional space. Such properties on curved surfaces, called intrinsic, have to be unaffected by a deformation without stress of the surface in the embedding space. Already in 1816 Gauss had engaged in geodetic problems, as Staeckel has shown in his article "Gauss als Geometer." [52] But his interest was particularly focused on this subject when he was asked by the Hanoverian government to serve as scientific adviser in an extensive geodetic survey of Hanover. (Göttingen at that time was under the government of Hanover.) As a result of his mathematical investigations in connection with this survey, which was performed under his direction until 1825, he published two important papers.[53] These papers, and especially his "Disquisitiones circa superficies curvas," [54] published in 1827, broke new ground and became through the work of Riemann the foundation of modern mathematical investigations into the structure of space. Once again we see that, historically viewed, abstract theories of space owe their existence to the practice of geodetic work, just as ancient geometry originated in the practical need of land surveying.

[52] Gauss, *Werke*, vol. 10, 2 Abh. IV.
[53] "Bestimmung des Breitenunterschiedes zwischen den Sternwarten von Göttingen und Altona" (1828; *Werke*, vol. 9); "Untersuchungen über Gegenstände der höheren Geodäsie" (1843, *Werke*, vol. 4).
[54] Gauss, *Werke*, vol. 4.

The theory of surfaces as such was not a new subject. Euler, Lagrange, and Monge already had investigated geometric properties on certain types of curved surfaces. But it was left for Gauss to study the problem in its generality and to lay thereby the foundations for differential geometry. In his "Disquisitiones" (already mentioned above), his first systematic exposition of quadratic differential forms, he investigated the possibility of an intrinsic determination of the curvature of a surface (today called the "Gaussian curvature"). Gauss's great contribution to differential geometry rests in his proof that this curvature, which is determined as the reciprocal product of the two principal radii, can be expressed in terms of intrinsic properties of the surface.

For this purpose Gauss assumed two families of curves to be drawn on the surface. Along each curve of the first family (the x_1-curves) x_2 is constant and along each of the other curves (the x_2-curves) x_1 is constant, much as in the ordinary Cartesian coördinate system along the ordinate y the value of the abscissa x is constant and vice versa. These curves are to cover the whole surface for varying values of the constants and any x_1 curve is to intersect any x_2 curve only in one single point. Any point P on the surface is consequently determined by the values of x_1 and x_2 of the two curves that intersect in it; x_1 and x_2 are called today the "Gaussian coördinates" of the point P on the surface. A familiar example is the system of longitude and colatitude coordinates on a sphere; here the x_2-curve with constant x_1 (longitude) is a meridian and the x_1-curve with constant x_2 (colatitude) is a latitude circle.

Now, if ds is the element of arc of a curve on the surface, it can be shown, using the Pythagorean theorem in Cartesian coordinates, that

$$ds^2 = g_{11}dx_1{}^2 + 2g_{12}dx_1dx_2 + g_{22}dx_2{}^2,$$

or, according to the familiar summation convention,

$$ds^2 = g_{mn}dx_mdx_n,$$

m and n to be summed over 1 and 2.

In this expression, as usual, the dx_m are infinitesimal increments of the Gaussian coördinates and the g_{mn} are magnitudes which in general depend on the Gaussian coördinates of the point in whose immediate vicinity the element of arc is to be computed.

In a continuous manifold of n dimensions its continuity and dimensionality do not yet allow us to infer any metrical properties, that is, properties to be ascertained by measurement. All that is known is that every point of the manifold is characterized by n numbers and that to closely adjacent points closely adjacent numbers correspond. But how can the distance between two given points be computed if only their coördinates are known? In axiomatic geometry the notion of congruency lies at the basis of measurements of distance or length. In practical geometry, however, with which physics is concerned, distance must be referred to the physical properties of a rigid body that can be transported from one place to another without change. To be sure, the rigid scale can be of arbitrary smallness. These ideas induced Riemann to see in ds, as used by Gauss in his surface theory, the appropriate mathematical expression for an infinitesimal element of length. Riemann thus assumed for a line element in a manifold of n dimensions with the general coördinates x_1, $x_2, \ldots, x_n$ the formula

$$ds^2 = g_{\mu\nu} \, dx_\mu \, dx_\nu,$$

μ and ν to be summed over $1, 2, \ldots, n$, and investigated the problem how to explore the geometry of this space on the basis of this expression.

This differential expression representing ds is usually called today the metric form or fundamental form of the space under consideration and the $g_{\mu\nu}$, owing to the invariance of ds, are the components of a covariant tensor of the second rank, the so-called fundamental tensor. A continuous n-dimensional manifold is called a Riemannian space, if there is given in it a fundamental tensor.

For the sake of historical accurateness we have to note that

Riemann apparently assumed that the concept of distance is intrinsic in space. Modern mathematics has shown that a logically consistent theory of a non-Riemannian space (that is, nonmetrical space) can be advanced in which the notion of distance is never encountered. The space of physical experience lends itself to measurements of length or distance, but it should be borne in mind that the concept of length or distance is foreign to the amorphous continuous manifold and has to be put in or "impressed" from without. For "length" and "distance" are operational concepts which find their mathematical counterpart through epistemic correlations. As will be explained on page 165, the element of length ds, as a mathematical invariant in Riemannian space, will be made to correspond to an infinitesimal "stretch" on a "practically rigid body."

In postulating the above formula Riemann showed that it provides sufficient, although not necessary, specifications for a line element satisfying the fundamental requirements of a distance function. The position of a point P is determined by n numbers $x_1, x_2, \ldots, x_n$. If $x_1 + dx_1, x_2 + dx_2, \ldots, x_n + dx_n$ denote the values of the coördinates of an adjacent (neighboring) point P', the length ds of the line element PP' must be expressed as a certain function of the increments $dx_1, dx_2, \ldots, dx_n$. If these increments are all increased in the same ratio, ds must also be increased in this ratio. If all the increments are changed in sign, the value of ds must remain unaltered. Assuming a simple algebraic relation between ds and the increments, these conditions suggest that ds must be an even root, the square root, fourth root, . . . of a positive homogeneous function of the $dx_1, dx_2, \ldots, dx_n$ of the second, fourth, . . . degree. Riemann selected the simplest hypothesis, namely, that ds is the square root of a homogeneous function of the increments of the second degree. He was fully aware of the arbitrariness in his determination of the length of the line element and emphasized the possibility of other expressions, as, for instance, the fourth power of ds as a biquadratic form of the coördinate differentials. The problem is of course connected with the question of the validity of the Pythagorean theorem in the vicinity of a point. Helmholtz,[55] Sophus Lie, and Weyl[56] attempted to show the necessity of assuming a quadratic form for the square of the line element. H. P. Robertson's investigations are also relevant to this problem.

Hermann von Helmholtz began his research on the structure of physical space because of his interest in the physiological problem of the localization of objects in the field of vision. In order to solve the problem of the dependence of ds on the increments dx_n, Helmholtz advances the principle of free mobility of rigid bodies and the principle of monodromy, that is, the assumption that a body being rotated about any arbitrary axis returns unchanged to its original position. It can readily be shown that the notion of congruence which lies at the basis of Helmholtz's principles imposes severe limitations on the a priori determination of the mathematical dependence of ds on the increments. Let there be given five points, A, B, C, D, E in three-dimensional space with the respective coordinates $x_A{}^1$, $x_A{}^2$, $x_A{}^3$, $x_B{}^1, \ldots, x_E{}^3$. The distance between any two points out of these five is given by a certain distance function, the variables of which are the corresponding coördinates of the two points involved. Let us now try to construct a congruent figure composed of five points A', B', C', D', E', in which the distance between any pair of points equals the distance between the corresponding pair of points in the original figure. Clearly, A' can be chosen arbitrarily in space; B', however, is then restricted to a certain surface, since its coördinates have to satisfy one equation; C' has to lie on a curve, its coördinates being submitted to two conditions; D' and E' are completely determined since their distances from A', B',

[55] Helmholtz, "Ueber die Tatsachen, die der Geometrie zu Grunde liegen." Cf. F. Lenzen, "Helmholtz's theory of knowledge," in *Studies and essays in the history of science and learning, offered in homage to George Sarton* (Schuman, New York, 1946), p. 309.
[56] H. Weyl, "Die Einzigkeit der Pythagoräischen Massbestimmung," *Math. Zeit.* 12, 114 (1922).

and C' are given. The assumption that $D'E'$ equals DE imposes a restriction on the mathematical formulation of the distance-function.

In general, n points in three-dimensional space have $3n$ co-ordinates and $n(n-1)/2$ mutual distances. We thus have $n(n-1)/2$ equations involving $3n$ coördinates, whereas the set of these n points, if considered as a rigid body, is determined by 6 parameters (degrees of freedom). The $3n$ coördinates may now be eliminated from the $n(n-1)/2$ equations and $\frac{1}{2}n(n-1) - 3n + 6 = \frac{1}{2}(n-3)(n-4)$ conditions will result.

Helmholtz's investigations found their strict mathematical elaboration in the works of Marius Sophus Lie.[57] Lie replaced the concept of mobility in space by the mathematical notion of a transformation between two systems of coördinates and reduced the geometric concept of congruence to the requirement of a certain invariance under such transformations. The displacement of a rigid body becomes equivalent to a one-to-one transformation of all space into itself, two successive displacements being replaceable by a third single transformation. Lie's theory of continuous groups, apart from its importance for axiomatic geometry (by showing that "congruence" is capable of a definition in terms of other fundamental geometric notions), has demonstrated that metric geometry is but the theory of the properties of certain particular congruence groups. Without having recourse to Helmholtz's assumption of monodromy, Lie comes to the conclusion that the only possible types of metric geometry are Euclidean, hyperbolic, and elliptic, a result which again imposes severe restrictions on the expression for ds.

Before concluding this digression on geometric investigations of the structure of space and resuming the subject of the further development of Riemann's contribution to the problem of space, a question will be asked that certainly deserves our attention,

---

[57] M. S. Lie, *Theorie der Tranformationsgruppen* (Leipzig, 1888–1893). See also Lie's *Ueber die Grundlagen der Geometrie* (Leipzig, 1890).

for it logically precedes all inquiries concerning the form of ds: How is it possible to define coördinates of points in space at all, as long as the concept of congruence is not yet determined? Von Staudt, whom Klein calls "einen der am tiefsten eindringenden Geometer, die je gelebt haben," thought he had found the solution in his *Geometry of position*[58] by using a repeated application of the quadrilateral construction for a harmonic range in the establishment of a system of projective coördinates independent of distance.

A similar attempt was made by Arthur Cayley by means of his projective equivalent of metric distance, employing the concept of cross ratio and leading finally to a vicious circle. As a matter of fact, the nature of these highly technical problems was fully understood only within the last five decades and interfered little with the early investigations concerning the structure of Riemannian space. Riemann assumed the validity of the Pythagorean theorem in the infinitely small. His theory of space, therefore, rests on geometric assumptions about infinitesimally small magnitudes. Being essentially a geometry of infinitely near points, it conforms to the Leibnizian idea of the continuity principle, according to which all laws are to be formulated as field laws and not by actions at a distance.

Riemann's geometry, in this respect, contrasted with the finite geometry of Euclid, can be compared with Faraday's field interpretation of electrical phenomena that formerly had been explained by actions at a distance. Weyl characterizes this situation as follows: "The principle of gaining knowledge of the external world from the behaviour of its infinitesimal parts is the mainspring of the theory of knowledge in infinitesimal physics as in Riemann's geometry."[59]

[58] Georg Karl Christian von Staudt, *Geometrie der Lage* (Nürnberg, 1847). See also von Staudt's *Beiträge zur Geometrie der Lage* (Nürnberg, 1856–1860).

[59] H. Weyl, *Space-time-matter* (trans. by H. L. Brose; London, 1922), p. 92.

"Straight lines" of Euclidean geometry are generalized in Riemannian space to "geodesic lines," [60] or simply "geodesics," to wit, lines of extreme distances between their terminal points. These geodesics, whose equations contain the components of the covariant fundamental tensor and their derivatives in certain definite combinations (Christoffel's symbols of the second kind), form a natural network throughout the n-dimensional manifold and can be used as a basis for the determination of its curvature. At a given point in the manifold let two infinitesimal vectors be given and the pencil of vectors linearly dependent upon them. With these vectors as initial elements, geodesics can be drawn, originating at the given point and generating a two-dimensional "geodesic surface" $\sigma^{\mu\nu}$ with its normal N. Riemann now defined the general curvature K_N of the n-dimensional manifold at the given point with respect to the normal N as the Gaussian curvature of this geodesic surface. It is obvious that the Riemannian curvature K_N depends on the orientation N of the geodesic surface and varies also from point to point. In other words, it is a measure both for the anisotropy and for the heterogeneity of space.

Now let the geodesic surface be represented by the oriented surface element $\sigma^{\mu\nu}$, an antisymmetric tensor of the second rank. It can be proved that the Riemannian curvature is then given by the expression

$$K_N = \frac{(\alpha\beta, \gamma\delta)\, \sigma^{\alpha\beta}\, \sigma^{\gamma\delta}}{(g_{\alpha\gamma}\, g_{\beta\delta} - g_{\alpha\delta}\, g_{\beta\gamma})\, \sigma^{\alpha\beta}\, \sigma^{\gamma\delta}}, \quad (\alpha < \beta, \gamma < \delta)$$

in which $(\alpha\beta,\gamma\delta)$ is the Riemann four-index symbol. It is related to the Riemann-Christoffel tensor by the equation

$$(\alpha\beta, \gamma\delta) = g_{\beta\epsilon} R^{\epsilon}_{\alpha\delta\gamma},$$

[60] As Klein remarks in his *Vorlesungen über die Entwicklung der Mathematik im 19. Jahrhundert* (Chelsea Publishing Company, New York, 1950), vol. 2, p. 148, with reference to Staeckel's *Zur Geschichte der geodätischen Linien*, the name "geodesic" as a technical term became common usage paradoxically only with Liouville (1850), that is, at a time when theoretical geometers were already not interested in practical geodesy.

with the usual designation for the components of the Riemann-Christoffel tensor.[61] Consequently,

$$(\alpha\beta, \gamma\delta) = R_{\alpha\beta\gamma\delta}$$

and

$$K_N = \frac{R_{\alpha\beta\gamma\delta}\, \sigma^{\alpha\beta}\sigma^{\gamma\delta}}{(g_{\alpha\beta}g_{\beta\delta} - g_{\alpha\delta}g_{\beta\gamma})\, \sigma^{\alpha\beta}\sigma^{\gamma\delta}}.$$

This equation shows clearly that the Riemannian curvature K_N vanishes everywhere if the Riemann-Christoffel tensor $R_{\alpha\beta\gamma\delta}$ is a constant zero tensor. Since the vanishing of the last-mentioned tensor is merely an analytic expression for the Euclidean structure of space, we infer that in Euclidean space the Riemannian curvature is everywhere equal to zero.

If the Riemannian curvature is independent of the orientation N of the geodesic surface element $\sigma^{\mu\nu}$ at every point in space, which certainly holds if space is isotropic throughout, then it is easy to show that

$$K(g_{\alpha\gamma}g_{\beta\delta} - g_{\alpha\delta}g_{\beta\gamma}) = R_{\alpha\beta\gamma\delta}.$$

Covariant differentiation of this equation and the use of Bianchi's identity leads to the result

$$\frac{\partial K}{\partial x^m} = 0.$$

In other words, the Riemannian curvature is a constant. According to this theorem, which was proved for the first time by F. Schur in 1886,[62] isotropy in every point of a Riemannian space implies its homogeneity.

Applying this result to physical space, which by the end of the nineteenth century was conceived to be isotropic as a matter of course, the following result was obtained: either (1) the Riemannian curvature is everywhere zero and space is Euclidean, or (2) it is a positive constant and space is spherical, or (3) it is

[61] See, for example, A. Einstein, *The meaning of relativity* (Princeton University Press, Princeton, 1953).

[62] F. Schur, "Räume konstanten Krümmungsmasses, II," *Math. Ann. 27* (1886).

a negative constant and space is hyperbolic. To sum up, only these three types of geometry are compatible with the isotropy of space. Riemann's complicated calculations seemed to have brought to light essentially nothing new.

Space is a form of phenomena, and, by being so, is necessarily homogeneous. It would appear from this that out of the rich abundance of possible geometries included in Riemann's conception only the three special cases mentioned come into consideration from the outset, and that all the others must be rejected without further examination as being of no account: *parturiunt montes, nascetur ridiculus mus!* [63]

Riemann, however, thought differently. He felt that labor was not lost. The assumption of a homogeneous space, in his view, does not take account of the existence of matter. Just as a strictly homogeneous magnetic or electrostatic field is never encountered in reality, so a homogeneous metrical field of space is only an idealization. Just as the physical structure of the magnetic or electrostatic field depends on the distribution of magnetic poles or electric charges, so the metrical structure of space is determined by the distribution of matter. With prophetic vision Riemann wrote: "The basis of metrical determination must be sought outside the manifold in the binding forces which act on it." [64]

These words were clearly an anticipation of some central ideas in Einstein's theory of gravitation, according to which the metrical structure, determined by the Einstein tensor $R_{\mu\nu}$, is related at every point of the space-time continuum to the mass-energy tensor $T_{\mu\nu}$ by the field equations

$$R_{\mu\nu} - \tfrac{1}{2}\, g_{\mu\nu}R = -\kappa T_{\mu\nu}.$$

The left-hand member of these famous equations involves the $g_{\mu\nu}$ and their derivatives, the right-hand member is an expression for the distribution of matter and energy, and κ is a constant related to the Newtonian constant of gravitation. The integration

[63] Weyl, *Space-time-matter*, p. 96.
[64] "Es muss also entweder das dem Raume zu Grunde liegende Wirkliche eine discrete Mannigfaltigkeit bilden, oder der Grund der Massverhältnisse ausserhalb, in darauf wirkenden bindenden Kräften, gesucht werden." Riemann, *Collected works*, p. 286.

of these field equations, in general no easy task, leads to the determination of the $g_{\mu\nu}$ as functions of the mass-energy distribution.

Riemann's anticipation of such a dependence of the metric on physical data seems to have been the logical solution of a dilemma to which the assumption of a variable curvature of space would have led. Since such curvature is an intrinsic property of space, that is, can be determined by geometric measurements within space itself, its very existence as a function of position would make it possible to designate position in space without resorting to a material coördinate system. Marks or labels could be assigned to points in space in accordance with the varying curvature, a process which could serve for their identification or distinction. In other words, absolute space would have been reinstalled. This difficulty is overcome by relating the local inhomogeneity to the material content of space.

Riemann's allusions were ignored by the majority of contemporary mathematicians and physicists. His investigations were deemed too speculative and theoretical to bear any relevance to physical space, the space of experience. The only one who allied himself firmly to Riemann was the translator of his works into English, William Kingdon Clifford. Moreover, already in 1870 Clifford saw in Riemann's conception of space the possibility for a fusion of geometry with physics. For Riemann, matter was the *causa efficiens* of spatial structure. By identifying cause and effect, a methodological procedure often encountered in the history of science, Clifford conceived matter and its motion as a manifestation of the varying curvature. He assumed that the Riemannian curvature as a function of time may give rise to changes in the metric of the field after the manner of a wave, thus causing ripples that may be interpreted phenomenally as motion of matter.

We may conceive our space to have everywhere a nearly uniform curvature, but that slight variations of the curvature may occur from point to point, and themselves vary with the time. These variations of the curvature with the time may produce effects which we not unnaturally attribute to physical causes independent of the geometry of our space. We might even go so far as to assign to this variation of the curvature of space "what really happens in that phenomenon which we term the motion of matter." [65]

Clifford's work from which this quotation has been taken was published posthumously by Karl Pearson, who states in the preface that the chapters on space and motion were dictated by Clifford himself in 1875. In 1876 Clifford published a paper "On the Space-Theory of Matter" in which he expressed similar ideas. He wrote:

I hold in fact
(1) That small portions of space are in fact of a nature analogous to little hills on a surface which is on the average flat; namely, that the ordinary laws of geometry are not valid in them.
(2) That this property of being curved or distorted is continually being passed on from one portion of space to another after the manner of a wave.
(3) That this variation of the curvature of space is what really happens in that phenomenon which we call the motion of matter, whether ponderable or etherial.
(4) That in the physical world nothing else takes place but this variation, subject (possibly) to the law of continuity.[66]

Clifford's suggestion, today no longer deemed as fantastic as in his own day, is in a certain sense the climax of a long development. For Aristotle, space was an accident of substance; for Clifford, so to speak, substance is an accident of space. The concept of space, after its emancipation during the Renaissance, seized totalitarian power in a triumphant victory over the other concepts in theoretical physics.

These speculations aroused great opposition among academic philosophers who still adhered to the Kantian doctrine according to which the axioms of Euclidean geometry were a priori judgments transcending reason and experience. In addition to such philosophical considerations, the advocates of Euclidean geome-

[65] W. K. Clifford, *The common sense of the exact sciences* (ed. by J. R. Newman; Knopf, New York, 1946), p. 202.
[66] *Proceedings of the Cambridge Philosophical Society* (1876).

try used the important argument that Euclidean geometry, in opposition to elliptic and hyperbolic geometry, was independent of any absolute length. This idea was strongly emphasized in particular by A. Gerstel, E. König, J. Cohn, K. Geissler, and H. Cornelius.[67] But marching along with truth is error. Thus it was maintained by many of these propounders of the impossibility of a non-Euclidean structure of physical space that only Euclidean space was homogeneous,[68] or it was erroneously asserted that non-Euclidean geometry presupposes Euclidean geometry.[69]

As late as 1900 the possibility of exploring by observation whether space is "Euclidean" ("flat") or "curved" attracted the attention of distinguished scientists. Thus K. Schwarzschild [70] published at the turn of the century a paper "On the admissible curvature of space" in which he tried to find an upper limit of the curvature of space (or its absolute value) without committing himself on the question whether physical space, if curved, is elliptic (curvature > 0) or hyperbolic (curvature < 0). On the basis of parallax statistics, taking into account the possible

[67] Adolf Gerstel, "Ueber die Axiome der Geometrie," *Beilage zum 16. Jahresbericht der philosophischen Gesellschaft, Wien* (1903), pp. 97–111. Edmund König, "Kant und die Naturwissenschaften," *Die Wissenschaft*, part 22 (1907). Jonas Cohn, "Voraussetzungen und Ziele der Erkenntnis," *Untersuchungen über die Grundfragen der Logik* (Leipzig, 1908). Kurt Geissler, *Moderne Verirrungen auf philosophisch-mathematischen Gebieten. Kritische und selbstgehende Untersuchungen* (1909); cf. K. Geissler, *Philosophie der Mathematik* (Interlaken, 1933). Hans Cornelius, *Grundlagen der Erkenntnistheorie. Transzendentale Systematik* (Munich, 1916; ed. 2, 1926).
[68] Cf. R. H. Lotze, *Grundzüge der Metaphysik* (Leipzig, 1884). Paul Natorp, *Die logischen Grundlagen der exakten Wissenschaften* (Leipzig, 1910). Hans Driesch, *Ordnungslehre, ein System des nicht-metaphysischen Teiles der Philosophie* (Jena, 1912).
[69] Cf. Alois Riehl, *Der philosophische Kritizismus und seine Bedeutung für die positive Wissenschaft* (Leipzig, 1879); Christoph von Sigwart, *Logik* (Freiburg, 1893), vol. 2, "Methodenlehre"; *Logic* (trans. by Helen Dendy (London, 1895), vol. 2; also H. Cornelius, *Transzendentale Systematik,* and H. Driesch, *Ordnungslehre*. For further literature on these erroneous assumptions, see the bibliography of R. Carnap, "Der Raum," *Kantstudien,* Ergänzungsheft No. 56 (1922).
[70] K. Schwarzschild, "Ueber das zulässige Krümmungsmass des Raumes," *Vierteljahrschrift der astronomischen Gesellschaft,* vol. 35 (1900), p. 337.

errors of observation, he came to the conclusion that, if space is hyperbolic, its radius of curvature is at least 64 light-years; on the assumption of an elliptic structure, the radius of curvature turned out to be at least 1600 light-years.

It was only toward the turn of the century that Poincaré demonstrated once for all the futility of this controversy and the fallacy of any attempt to discover by experiment which of the mutually exclusive geometries applies to real space. Measurement, he insists, is never of space itself, but always of empirically given physical objects in space, whether rigid rods or light rays. Regarding the structure of space as such, experiment can tell us nothing; it can tell us only of the relations that hold among material objects. Suppose, Poincaré says, a deviation from two right angles had occurred in the triangulation carried out by Gauss, would this necessarily have constituted a refutation of Euclidean geometry? For there would be nothing to prevent us from continuing to use Euclidean geometry on the assumption that light rays are curved. Nothing could disprove such an assumption. So the highly important conclusion emerges that experience can neither confirm nor refute a geometry, whichever geometry it be. What geometry one chooses is, for Poincaré, merely a matter of convenience, a convention. We select that system of geometry which enables us to formulate the laws of nature in the simplest way.

Poincaré was convinced, on the basis of this conclusion, that Euclidean geometry, the familiar abstraction from common experience with solid bodies and light rays, would always remain the favored system. He was wrong, as the development of general relativity has shown.[71] The classical example of gravitation will suffice to make this point clear.

Gravitation, as understood by the theory of general relativity,

[71] He was wrong also as far as the logical simplicity of Euclidean geometry is concerned. As modern research has shown, Euclidean geometry lacks the distinction of logical simplicity and "hyperbolic geometry is the only one which can be developed from a few simple assumptions concerning joining, intersecting, and continuity alone" (K. Menger).

is to be comprehended in the geometric structure of space-time. This fusion not only made physical theory logically more unified and simpler, but led to the great triumph of the new theory over classical physics (in the famous observable effects: advance of the perihelion of Mercury, deflection of light rays in a gravitational field, etc.). Let the geometry in a coördinate system (x^1, x^2, x^3, x^4) be determined by the field equations and have the line element

$$ds^2 = g_{\mu\nu}\, dx^\mu\, dx^\nu.$$

If an observer, following Poincaré's suggestion, adheres to the Euclidean (or, in a four-dimensional continuum, to the so-called Galilean) metric and selects as his line element

$$ds^2 = -(dx^1)^2 - (dx^2)^2 - (dx^3)^2 + (dx^4)^2,$$

he will soon realize that his ds cannot be made compatible with his observational results. Thus, a freely moving particle will not follow the path described by the condition

$$\delta \int ds = 0.$$

If our observer is not willing to revise his geometry and to change its metric, he will be led to the conclusion that the particle, although apparently undisturbed, deviates from the geodesic line of his geometry, that is, from uniform motion in a straight line. This clearly contradicts the Galilean principle of inertia. In order to remove this contradiction he will suppose the existence of a "field of force" (for example, gravitation) and inquire into its physical properties, without realizing that this "field of force" is but a fiction, invoked by the discrepancy between the appropriate "natural" geometry — as required by the field equations — and the Euclidean geometry to which he adheres. Our observer's predilection for his familiar geometry has led him to an enormous complication in his physical theory.

A system of pure axiomatic geometry does not suffice, if geometry is to be applied to the space of physics. What is needed is a correlation between the geometric concepts of the abstract system with physical objects or physical processes. As Einstein pointed out in a lecture[72] at the Berlin Academy of Sciences in 1921, the most natural and simple assumption would relate the physical behavior of rigid bodies to the geometric properties of solid bodies in Euclidean geometry. This, however, does not necessarily imply that the space of physics is Euclidean.

How such a correlation can be established is explained also in Einstein's "Physik and Realität." [73] The conceptual construction of the notion of space in modern physics is based on the empirical fact, already noted by Poincaré, that there exist two kinds of alteration of physical objects, changes of state and changes of position. In contrast to the former, it is the latter type of change that can be reversed by the arbitrary motions of our bodies. "That there are bodily objects to which we have to ascribe within a certain sphere of perception no alteration of state, but only alterations of position, is a fact of fundamental importance for the formation of the concept of space (in a certain degree even for the justification of the notion of the bodily object itself)." Such a bodily object is called by Einstein "practically rigid." The position of either of two given practically rigid bodies can be changed without changing the position of the pair as such. So we get the concept of "relative position," a special case of which is "contact" of two bodies at a point. Any two points on a practically rigid body define a "stretch" (as Lenzen calls it in his article on "Einstein's theory of knowledge" [74]). Two stretches on two practically rigid bodies, one on each, are defined as equal if the points of one are in contact ("coincide") with the corresponding points of the other. We have now to

[72] Later published under the title *Geometrie und Erfahrung* (Erweiterte Fassung des Festvortrages, gehalten an der Preussischen Akademie, Berlin, 1921). Cf. *Sidelights on Relativity* (trans. by G. B. Jefferey and W. Perrett; London, 1922).

[73] Published in the *Journal of the Franklin Institute 221*, 313–347 (1936); English translation by J. Piccard, pp. 349–382. The following outline is based primarily on this article.

[74] In P. A. Schilpp, ed., *Albert Einstein, Philosopher-scientist* (Tudor, New York, 1950), p. 355.

postulate that two stretches, once determined to be equal, are always and everywhere equal. This relation of equality, a symmetric, reflexive, transitive relation, independent of position and time, can now be correlated with the abstract notion of congruence in Euclidean geometry.

To be sure, experience can show equality between two stretches only if they are adjacent; in addition, it can show that two stretches found to be equal remain so when separated and then brought together again. But experience can tell us nothing about equality between two nonadjacent stretches. The relation of equality "at a distance," so to speak, is a postulating generalization of the original definition of equality. The postulate is essentially only a positive expression for the rejection of the concept of absolute space. Were absolute space assumed to exist, it would serve as a gage by which the validity of the "postulate" could be tested.

The postulate also deprives the famous space fantasies, popularized by Delboeuf [75] and especially Poincaré,[76] of all physical meaning. Poincaré imagines the case of a uniform expansion of the universe. All its dimensions increase over night a thousand times. What was formerly one meter, will now measure one kilometer. Clearly, such an expansion is beyond all physical verification, for whatever the measuring instrument employed, it too will have increased in the same ratio. Even if the imagined expansion were not uniform, but anisotropic, say ten times as much in one direction as in another direction perpendicular to the first, it too would go unnoticed by any observer. The fact that our postulate precludes such an expansion of the universe is equivalent to the statement that such an expansion can be formulated only on the assumption of the existence of absolute space.

[75] J. R. L. Delboeuf, *Prolégomènes philosophiques de la géometrie* (Paris, 1860), containing the substance of his lectures before the Royal Belgian Academy of Sciences, "Nains et géants" and "Mégamicros ou les effets sensibles d'une réduction proportionelle des-dimensions de l'univers." See also his articles "L'Ancienne et les nouvelles géométries," *Revue philosophique 36*, 449 (1893).
[76] Poincaré, *Science and method* (Dover, New York, 1952), p. 94.

From the historical point of view it is interesting to note that Delboeuf and Poincaré were not the first who stated that a uniform expansion or contraction of all magnitudes in the universe would be unobservable. Laplace[77] had pointed out in 1808 that on the basis of Newtonian physics an expansion of all distances, accelerations, and masses in the same ration would have no effect on the order of physical events: the behavior of the physical universe is independent of the absolute magnitude of the scale. Laplace did not mention these considerations because of their relevance to the problem of space, but rather to show the unique importance of Newton's law of gravitation, which for the end of the eighteenth century was the archetype of all physical laws. Only the inverse-square law of force seemed to be compatible with the complete independence of the absolute magnitude of the scale. Let us give a simple example. The attraction of the sun (a sphere of radius R) on the earth at a distance D is proportional to R^3/D^2. Under the influence of this attraction the earth moves in unit time a distance A in a centripetal direction. Now assume that R, D, and A all increase in the ratio $n:1$. The force of attraction is now proportional to $(nR)^3/(nD)^2 = n(R^3/D^2)$, that is, it is increased n times, and it stands to the increased centripetal distance nA in the same relation as before.

Laplace's explications, one is tempted to say, may serve as an a priori demonstration of the validity of Newton's law of gravitation, once the independence of an absolute scale for all physical dimensions is granted. Incidentally, Delboeuf tried to prove, along similar lines, the validity of Euclid's axiom of parallels on the assumption of the homogeneity of space.[78]

So it comes about that the physical concept of equality is now made to correspond to the mathematical notion of congruence, just as the physical behavior of practically rigid bodies is made

[77] P. S. Laplace, *Exposition du système du monde* (Paris, 1808), book 4, chap. 16.
[78] L. Couturat, "Note sur la géométrie non-euclidienne et la relativité de l'espace," *Revue de métaphysique et de morale* 1, 302 (1893).

to correspond to the mathematical properties of solid bodies in Euclidean geometry.

It will be noticed that this operational or epistemic correlation between the physical behavior of practically rigid bodies and the purely abstract notion of congruency in deductive geometry holds whether axiomatic geometry adopts for its systematic construction the notion of movement through space or not. Most modern treatments of axiomatic geometry (for example, Hilbert's *Foundations of geometry*) eliminate this notion by a skillful adoption of a congruence axiom.[79] This procedure, most probably, has its historical origin in the predominant influence of Kantian critical philosophy on European, and in particular German, thought in the nineteenth century. For Kant's transcendental aesthetic comprises only the elements of space and time, whereas the concept of motion, not known a priori but only by experience, belongs to the realm of sensibility. Its introduction as a primary notion into deductive geometry seemed accordingly to violate the a priori synthetic character of the science of space.

These considerations are only of historical importance for the changed attitude of the twentieth century. Nevertheless, it is satisfying to realize that Einstein's procedure is unaffected by this problem, although an inclusion of movement among the primary notions of geometry could perhaps lead to a certain simplification. More important, however, is E. A. Milne's criticism concerning the postulated invariance of the length of rigid bodies under transport. Milne argued that such a statement as "a rigid measuring rod is unaltered under transport" is void of any operational specification, since no standard of length is available at the new position of the rod. In his manuscript for the Edward Cadbury Lectures, which he was to deliver at the University of Birmingham in 1950, Milne wrote:

It is part of the debt we owe to Einstein to recognize that only "operational" definitions are of any significance in science: we must be in a position to state a test by which we can tell whether a given

[79] See Felix Klein, *Elementary mathematics from an advanced standpoint* (Dover, New York, 1939), vol. 2, "Geometry," p. 175.

entity may be identified with one mentioned in the definition. A definition, in other words, must be couched in terms of "observables." Einstein carried out his own procedure completely when he analysed the previously undefined concept of simultaneity, replacing it by tests using the measurements which have actually to be employed to recognize whether two distant events are or are not simultaneous. But he abandoned his own procedure when he retained the indefinable concept of the length of a "rigid" body, i.e., a length unaltered under transport. The two indefinable concepts of the transportable rigid body and of the simultaneity are on exactly the same footing; they are fog-centres, inhibiting further vision, until analysed and shown to be equivalent to conventions.[80]

In his attempt to find an operational meaning for "permanence in length under transport" Milne adopted a method of reducing length to time determination. His procedure can be compared in principle with the well-known radar technique for measuring distances of remote objects. With clocks having suitable graduations,[81] coördinates can be assigned to various observers or objects. In particular, the distance r of observer B from observer A can be defined by taking an appropriate combination of clock readings t_1 and t_3 — corresponding to the epochs of transmission and reception of light signals — yielding $r = \frac{1}{2}c(t_3 - t_1)$, where c is a universal constant that later turns out to be the velocity of light. In general, "distances" and "lengths" (differences of distance coördinates) become definite for an observer equipped with a graduated clock. Now, if a measuring rod, originally at rest relative to such an observer is transported to a new position of rest relative to the same observer, and if its "lengths" (as measured by the radar technique) in the two positions are equal, then "the rod is said to have undergone a rigid body displacement by this clock." Thus, according to Milne, a meaning of "permanence of length under transport" depends on the provision of graduated clocks. In contrast to Einstein, and perhaps

[80] Milne's Cadbury Lectures were published posthumously under the title *Modern cosmology and the Christian idea of God* (Clarendon Press, Oxford, 1952); the quotation is from p. 35.

[81] For details see E. A. Milne, *Kinematic relativity* (Clarendon Press, Oxford, 1948).

even more in contrast to Eddington, Milne regarded the measurement of time as much more fundamental than the measurement
of length. By reducing the determination of length to the measurement of time Milne tried to dispense with rigid measuring
rods.

> The reason why it is more fundamental to use clocks alone rather
> than both clocks and scales or than scales alone is that the concept
> of the clock is more elementary than the concept of the scale. The
> concept of the clock is connected with the concept of "two times at
> the same place," whilst the concept of the scale is connected with the
> concept of "two places at the same time." But the concept of "two
> places at the same time" involves a convention of simultaneity, namely,
> simultaneous events at the two places, but the concept of "two times at
> the same place" involves no convention; it only involves the existence
> of an ego.[82]

Milne's analysis of the "permanence of length under transport"
seems to be justified from the purely logical point of view. As
far as its scientific fruitfulness is concerned, it certainly cannot
be compared with Einstein's analysis of the concept of "simultaneity." Einstein's critique of the traditional notion of simultaneity
led to a radical revision of classical conceptions, whereas Milne's
important contributions to theoretical physics are certainly not
primarily the result of his critique of the "permanence of length
under transport."

It was Einstein who made it clear that geometry, when applied
in this way to the exploration of physical space, ceases to be an
axiomatic deductive science and becomes one of the natural
sciences, indeed, the oldest of all. Poincaré was only partly right:
it is a matter of convention which geometry we adopt, but only
as long as no assumptions are made concerning the behavior of
physical bodies as implied in the measurements. Once these
assumptions are laid down, the choice of the geometric system is
determined. As Einstein explains, it is the sum total of the assumptions of correlation and of the system of abstract geometry
that has to conform to experience. Once the principle that relates

[82] Milne, *Modern cosmology and the Christian idea of God,* p. 46.

rigid bodies to Euclidean solids is accepted, it is experience that conditions the choice of geometry. For example, in a reference system that rotates relatively to an inertial system, the laws of placing[83] rigid bodies no longer correspond, owing to the Lorentz transformation, to the rules of Euclidean geometry. In accordance with our fundamental postulate, only one choice is possible and Euclidean geometry must give way to Riemannian geometry. Hence it is clear that the structure of the space of physics is not, in the last analysis, anything given in nature or independent of human thought. It is a function of our conceptual scheme.

Space as conceived by Newton proved to be an illusion, although for practical purposes a very fruitful illusion — indeed, so fruitful that the concepts of absolute space and absolute time will ever remain the background of our daily experience. Acknowledging this fact, Einstein writes in his "Autobiographical notes":

Newton forgive me; you found the only way which, in your age, was just about possible for a man of highest thought and creative power. The concepts which you created, are even today still guiding our thinking in physics, although we now know that they will have to be replaced by others farther removed from the sphere of immediate experience, if we aim at a profounder understanding of relationships.[84]

Thus Newton's conception of absolute space and its equivalent, Lorentz's ether, were shorn of their ability to define a reference system for the measurement of velocities. This was accomplished by the special theory of relativity. Within the framework of this theory, however, space as such was still a basic concept. To be sure, as part of the four-dimensional Minkowski space-time continuum it certainly had lost any individual distinction; an infinite number of coördinate systems were physically equivalent. It had, however, its own representation as an inertial system. Owing to the relativization of simultaneity, furthermore, the notion of action at a distance had to be discarded and the adoption of the field concept as the basic element of the theory had been suggested. This program was carried through by the general theory

[83] For example, along a circle whose center lies on the axis of rotation.
[84] In P. A. Schilpp, ed., *Albert Einstein, Philosopher-scientist,* p. 31.

of relativity, whereby the inertial system was replaced by the displacement field, a component part of the total field,

this total field being the only means of description of the real world. The space aspect of real things is then completely represented by a field, which depends on four coördinate-parameters; it is a quality of this field. If we think of the field as being removed, there is no "space" which remains, since space does not have an independent existence.[85]

In the conceptual construction of space according to modern science, the three-dimensionality of space or the four-dimensionality of the space-time continuum appears as an accidental feature, justified only by experience. For three numbers or coordinates suffice to locate a "point-object" in space and four coördinates determine a "point-event" in space-time unambiguously. That the three-dimensionality of space has to be accepted as accidental was considered already in antiquity as a serious flaw and an essential deficiency in a deductive theory of space. Aristotle considered this problem worthy of a detailed discussion in the opening chapter[86] of his *De caelo* and thought he could solve it in the spirit of Pythagorean mythical notions of perfection. Aristotle's arguments are recapitulated in the discussion between Salviatus and Simplicius in Galilei's *Dialogue on the great world systems*. Says Simplicius with reference to Aristotle:

Do you not have there the proof that there are no more than three dimensions, because those three are all things and are everywhere? And is this not confirmed by the doctrine and authority of the Pythagoreans, who say that all things are determined by three, beginning, middle, and end, which is the number of All? And where do you leave that argument, namely, that, as it were by the law of Nature, this number is used in the sacrifices of the gods? And why, being dictated by Nature, do we attribute to those things that are three, and not to less, the title of all? . . . Moreover, in the fourth text, does he not, after some other doctrines, prove it by another demonstration, viz. that no transition is made but according to some defect (and so there is a transition or passing from the line to the surface, because the line is defective in breadth) and that it is impossible for the per-

[85] Albert Einstein, *Generalization of gravitation theory*, a reprint of Appendix II from the fourth edition of *The meaning of relativity* (Princeton University Press, Princeton, 1953), p. 163.

[86] Aristotle, *De caelo*, 268 a.

fect to want anything, it being every way so; therefore, there is no transition from the Solid or Body to any other magnitude. Do you not think that by all these places he has sufficiently proved how there is no going beyond the three dimensions, Length, Breadth, and Thickness, and that therefore the body or solid, which has them all, is perfect? [87]

That such a demonstration does not satisfy the requirements of scientific rigor is Galilei's firm conviction. In fact, Salviatus replies:

To tell you true, I do not think myself bound by all these reasons to grant any more than this — that that which has beginning, middle, and end may be, and possibly ought to be, called perfect. But I cannot grant that, because beginning, middle, and end are three, the number three is a perfect number and has a faculty of conferring perfection on those things that have it. Neither do I understand nor believe that, for example, of feet, the number three is more perfect than four or two; nor do I conceive their number of four to be any imperfection in the elements; nor that they would be more perfect if they were three. Better therefore if he had left these subtleties to the rhetoricians and had proved his intent by necessary demonstration; for so it behooves to do in demonstrative sciences. [88]

As a matter of fact, Simplicius, the great commentator of Aristotle, referred to the insufficient demonstration in *De caelo* and compared it with Aristotle's theory of place as expounded in the *Physics*, where place (or space), in Simplicius's view, is essentially conceived as a two-dimensional extension. [88] Simplicius contrasts this latter treatment with the teachings of Strato, the Platonists, and the Stoics who stressed the three-dimensionality of space.

Apart from these scanty remarks, the three-dimensionality of space as a problem was scarcely discussed in antiquity or in the Middle Ages. Euclid's definition I in Book XI of the *Elements:* "A solid is that which has length, breadth, and depth," with the implicit identification of solids and bodies, was accepted without

[87] Galileo Galilei, *Dialogue on the great world systems* (trans. by T. Salusbury, ed. by Giorio de Santillana; University of Chicago Press, Chicago, 1953), p. 13. For the section of the proof omitted in this edition, see Emil Strauss's translation, *Dialog über die beiden hauptsächlichen Weltsysteme* (Leipzig, 1891), p. 12.
[88] *Ibid.*
[89] Simplicius, *Physics*, 601.

further questions. This seemed to be only natural since the
notions of surface, line, and point came later to be defined by
the process of abstraction from the concept of the solid. Moreover, the problem was dismissed by identifying three-dimensionality with body, as had been done already by Isaac Judaeus.[90]
Even Leibniz, who, as we have seen, submitted the concept of
space to a most critical analysis, took little notice of the problem
of the dimensionality of space. Recognizing that space has three
dimensions, he bases this statement on purely geometric considerations: "Le nombre ternaire est déterminé . . . par une ne-
cessité geómétrique: c'est parce que les Géomètres ont pu dé-
montrer qu'il n'y a que trois lignes droites perpendiculaires entre
elles, qui se puissent coupre dans un même point." [91]

With the rise of non-Euclidian geometry and other generalizations of classical geometry it became evident that pure mathematics, not logically confined to three dimensions, could operate
consistently with concepts of space that possesses any arbitrary
number of dimensions. The question why ordinary space possesses just three dimensions was considered from now on as a
problem in physics or in logic applied to real existence. Nevertheless, it was not always understood that a discussion which
proceeds wholly within the boundaries of geometric notions
necessarily cannot be decisive.[92]

One of the first for whom the three-dimensionality of space
became a problem of physics was Kant. Already in his *Gedanken
von der wahren Schätzung der lebendigen Kraft* he considers the
possibility of spaces having different dimensionalities.

Eine Wissenschaft von allen diesen möglichen Raumarten wäre
ohnfehlbar die höchste Geometrie, die ein endlicher Verstand unter-
nehmen könnte . . . Wenn es möglich ist, dass es Ausdehnungen von

[90] *Isaaci operi omnia lat.* (Leyden, 1515–16), "Liber de elementis."
[91] J. E. Erdmann, ed., *Leibnitii opera philosophica* (Berlin, 1860), p. 606.
[92] This applies, for example, to Whewell's demonstration in his *History of
scientific ideas* (London, 1858), vol. 1, p. 97, which is a variation of a
geometric proof given already in the beginning of the eighteenth century
by Leibniz in his *Essais de théodicée* (Amsterdam, 1710), §§351 and 196.

anderer Abmessung gebe, so ist es auch sehr wahrscheinlich, dass sie Gott wirklich irgendwo angebracht hat.[93]

However, he thought he had discovered the reason for the three-dimensionality of physical space of our experience in Newton's law of gravitation, according to which the intensity of the force decreases with the square of the distance.[94]

Gauss, in a letter to Gerling,[95] refers to a generalization of his considerations on symmetry and congruence for a geometry of more than three dimensions, "for which we human beings have no intuition, but which considered *in abstracto* is not inconsistent." As Sartorius von Waltershausen reports, Gauss considered the three-dimensionality of space not as an inherent quality of space, but as a specific peculiarity of the human soul.[96] Gauss seems to have understood that the question whether space has three dimensions or more stands on the same footing as the problem of the Euclidean or non-Euclidean character of space. Both questions need for a decision an external criterion, foreign to pure mathematics.

Certain ideas in Herbart's philosophy[97] seem to have had a great influence on Riemann and H. Grassman in their formulation of a manifold with an arbitrary number of dimensions. Grassmann's ingenious *Theory of extensions*,[98] published in 1844 and

[93] Kant, *Gedanken von der wahren Schätzung der lebendigen Kraft*, §10; see reference 16.

[94] For a detailed analysis of Kant's argument, see Ueberweg's similar demonstration on page 177. F. W. J. Schelling, in his *System des transcendentalen Idealismus* (Tübingen, 1800), pp. 176–185, also tries to explain the three-dimensionality of space, under the influence of Kant, by dynamical considerations. In addition to the Kantian-Newtonian forces of attraction and repulsion, Schelling postulates a third force (newly invented by him) and relates the dimensions of length, breadth, and depth to these three forces. According to Schelling, not only spatial three-dimensionality but also electricity and magnetism owe their origin to these forces.

[95] Letter of April 8, 1844.

[96] Sartorius von Waltershausen, "Gauss zum Gedenken," in Gauss, *Werke*, vol. 8, p. 268.

[97] Cf. J. F. Herbart, Habilitationsthese (October 23, 1802); cf. *Schriften zur Metaphysik* (Leipzig, 1851), part 2, chap. 4, "Vom Körperlichen Raume," p. 203.

[98] *Ausdehnungslehre* (1844; ed. 2, Leipzig, 1878).

in a second revised and amplified edition in 1862, was completely ignored at that time. Only the twentieth century began to understand the importance of Grassman's generalized algebraic treatment of n-dimensional spaces.

In mathematics proper, however, the nineteenth century was very successful in clarifying the concept of dimension, particularly after the development of the affine and projective geometry. J. Pluecker, generalizing the basic idea of the principle of duality in projective geometry, showed that the dimensionality of a space depends not only on topological properties, but also on the choice of the elements out of which space is constructed. Thus, for example, a Euclidean plane is three-dimensional if considered as a manifold of circles (two coördinates determine their centers and the third their radii). On the other hand, the plane appears as a five-dimensional manifold if conics are chosen as the basic space elements. In short, dimensionality in Pluecker's view is not an absolute attribute of space, but depends upon the basic elements that constitute the space. However, referring to Cartesian coördinates, it was shown by L. E. J. Brouwer in 1911 that the dimensionality of space is a topological invariant, that is to say, it remains invariant under any continuous transformation of the coördinates.

This short digression into the history of mathematics has been inserted because the mathematical development stimulated new interest among philosophers and physicists in the corresponding problem of the dimensionality of physical space. Indeed, it almost seems as if these mathematical generalizations of spatial dimensionality were looked upon as a challenge for scientists to prove the three-dimensionality of ordinary space. So numerous attempts were made in the course of the nineteenth century to prove that space in physics has only three dimensions. Remarkable because of its method is Bolzano's attempt.[99] He refers in his proof to

<hr />

[99] B. Bolzano, "Versuch einer objektiven Begründung der Lehre von den drei Dimensionen des Raumes," *Abhandlungen der böhmischen Gesellschaft der Wissenschaften* (Prague, 1843).

the temporal order of the immediately given contents in our consciousness and tries to show that the assumption of a three-dimensional space is indispensable for the construction of a correspondence between these contents and the external objective causal connectivity.

A very popular proof, based on similar lines, was that of Ueberweg.[100] Ueberweg deduced the reality of three-dimensional extension from the reality of time or temporal sequence, founded on internal experience. In his view, the empirically given order of time, the succession of day and night, of winter and summer, is based on mathematico-physical laws that presuppose three-dimensional space. The causal interconnection of physical processes demands three-dimensionality. Like Kant, Ueberweg mentions in this connection Newton's law of gravitation, according to which the intensity of gravitation between constant masses decreases with the square of the distance. This law, according to Ueberweg, presupposes a three-dimensional space. For in a space of only two dimensions the intensity would decrease with the distance itself (the circumference of the circle being proportional to the radius), in the space of three dimensions the intensity decreases with the square of the distance (the surface of the sphere being proportional to the square of the radius), and in a space with more dimensions the intensity would decrease with a higher power of the distance, in accordance with the condition that every point affected receives a proportional part of the total effect. In other words, since in an n-dimensional space the gravitational force must decrease according to the last-mentioned condition with the $(n-1)$th power of the distance, and since Newton's law established a decrease with the second power of the distance, n must be three. It is obvious that Ueberweg's proof fails if the a priori character of Newton's law is rejected. Moreover, even if the number of dimensions of the objective world should disagree with that of our phenomenal world, some mathematical order among the phenomena could still be conceived as possible, even if it

[100] F. Ueberweg, *System der Logik* (ed. 5, Bonn, 1882), p. 113.

were a distorted order or a kind of projected order. F. A. Lange discusses this possibility and says:

Astronomy is but a special case, for which, under other conditions, something else might be substituted. For the rest, we have no absolute standard as to what we might demand as regards the intelligibleness of the world, and for this reason alone Ueberweg's standpoint is really based upon a concealed *petitio principii*.[101]

Natorp's attempt to find an upper limit for the number of dimensions is based on his imposed restriction to find the minimum number of dimensions necessary and sufficient to guarantee a unique, closed, homogeneous, and continuous connection among spatial directions.[102] Through a consequent application of the notion of continuous rotation he assumes that he has achieved his purpose. Progressing from the straight line to the plane, to complete the manifold of directions, and progressing from the plane to three-dimensional space to complete the manifold of rotations, his deduction comes to an end apparently because the conception of a motion of a three-dimensional space as a whole lacks all intuition based on experience. It would certainly be beyond the scope of this chapter to discuss the other so numerous "proofs" of the three-dimensionality of space, as, for example, that given by Hegel[103] or that proposed by Trendelenburg.[104]

It is, however, curious to note that the idea of a fourth dimension was cordially welcomed in spiritual circles. Henry More had already applied this notion for his spiritualistic conception of what he called "*spissitudo essentialis*." In his *Enchiridion metaphysicum* he writes: "Ita ubicumque vel plures vel plus essentiae

[101] F. A. Lange, *History of materialism* (ed. 3, trans. by E. C. Thomas; Humanities Press, New York, 1950), book 2, "History of materialism since Kant," p. 226. For Ueberweg's reply to Lange's criticism, see Ueberweg, *Geschichte der Philosophie* (ed. 2, 1869), vol. 3, p. 303.

[102] Natorp, *Die logischen Grundlagen der exakten Wissenschaften*, p. 306. For a similar proof, see Friedrich Pietzker, "*Die dreifache Ausdehnung des Raumes*," *Unterr. — Bl. f. Mathem. u. Nat. 8*, 39 (1902).

[103] G. W. F. Hegel, *Encyklopädie der philosophischen Wissenschaften im Grundriss* (Leipzig, 1905), part 2, sec. 255, p. 214.

[104] F. A. Trendelenburg, *Logische Untersuchungen* (Leipzig, 1870), p. 226.

in aliquo ubi continetur quam quod amplitudinem huius adaequat, ibi cognoscatur quarta haec dimensio, quam apello spissitudinem essentialem." [105] Supernatural phenomena as provoked by spiritualists in their séances were accounted for on the assumption of a fourth dimension. Most famous in this respect are the experiments performed by the German professor of astronomy, J. K. F. Zöllner of Leipzig, in which many of his distinguished colleagues served as witnesses. Experiments of topological character, such as the untying of knots tied in closed loops of string or the famous instances known as "apports," the sudden appearance and approach of an object from nowhere, were explained as motions or processes in the fourth dimension of space. In his comprehensive work *Transcendental physics*[106] Zöllner tries to explain on the basis of this hypothetical dimension not only phenomena that occur in spiritualistic sittings but also religious miracles of all kinds. In his *Wissenschaftliche Abhandlungen*, of which the *Transcendental physics* forms the third volume, he refers to patristic theology (Hieronymus, Augustine, Cassiodorus, Gregory the Great)[107] and even to the teachings of his contemporary Mach,[108] as congenial to his theory of the fourth dimension. In his search for modern theological support for his theory of a four-dimensional space Zöllner refers to Friedrich Christoph Oetinger[109] and to Johann Ludwig Fricker,[110] Oetinger's friend. These two theologians used some vague unorthodox formulation,

[105] H. More, *Enchiridion metaphysicum*, I, 28, §7. See also Robert Zimmermann, *Henry More und die vierte Dimension des Raumes* (Vienna, 1881).

[106] J. K. F. Zöllner, *Transcendental physics* (trans. by C. C. Massey; London, 1880). The original *Transcendentale Physik*, vol. 3 of *Wissenschaftliche Abhandlungen* (Leipzig, 1878), was dedicated to William Crookes. See also Zöllner's article in *Quarterly Journal of Science* (April 1878) and G. E. Barnard, *The supernormal* (London, 1933), chap. 8.

[107] Zöllner, *Transcendentale Physik*, p. 600.

[108] *Ibid.* p. lxxxvii.

[109] *Des Wirttembergischen Prälaten Friedrich Christoph Oetinger sämmtliche Schriften* (ed. by K. C. E. Ehmann; Stuttgart, 1858).

[110] *Johann Ludwig Fricker, ein Lebensbild aus der Kirchengeschichte des 18. Jahrhunderts* (ed. by K. C. E. Ehmann; Heilbronn, 1872).

involving the notion of a fourth dimension, in their attempt to explain — in a quasi-geometric way — two Biblical passages:

Canst thou by searching find out God? canst thou find out the Almighty unto perfection?

It is as high as heaven; what canst thou do? deeper than hell; what canst thou know?

The measure thereof is longer than the earth, and broader than the sea.[111]

That Christ may dwell in your hearts by faith; that ye, being rooted and grounded in love,

May be able to comprehend with all saints what is the breadth, and length, and depth, and height;

And to know the love of Christ, which passeth knowledge, that ye might be filled with all the fulness of God.[112]

With regard to Christian teratology Zöllner says: "Das sacrificium intellectus welches die christlichen Wunder vom Verstande bisher verlangten, ist durch die Entdeckung jenes neuen Gebietes der Physik — der Transcendentalphysik — zum ungetrübten Genusse des Neuen Testamentes nicht mehr erforderlich." [113] That literature of this kind was popular not only at the end of the last century is shown today by the wide circulation of C. H. Hinton's books, *A new era of thought* [114] and *The fourth dimension,*[115] and in particular by P. D. Ouspensky's *Tertium organum, A key to the enigmas of the world.*[116]

Poincaré attempted to demonstrate the three-dimensionality of the space of experience by the following simple topological consideration.[117] Space cannot be separated into parts by isolated points (as in the case of a one-dimensional extension) nor by

[111] Job 11:7–9.
[112] Ephesians 3:17–19.
[113] Zöllner, *Wissenschaftliche Abhandlungen,* vol. 2, part 2, p. 1187. For a contemporary critique of Zöllner's theory, see Gutberlet, *Die neue Raumtheorie* (Mainz, 1882).
[114] New York, ed. 2, 1923. See also P. D. Ouspensky, *A new model of the universe* (Knopf, New York, 4th ptg., 1944), chap. 2, "The fourth dimension," pp. 61–100; G. B. Burch, "The philosophy of P. D. Ouspensky," *Review of Metaphysics* 5, 247 (1951).
[115] London, 1888.
[116] Allen and Unwin, London, 1934.
[117] H. Poincaré, *Dernières pensées* (Paris, 1917), p. 61.

curves (as in the case of two-dimensional extensions). Since, however, a closed surface separates space into disjunctive parts, Poincaré thought that he had found the fundamental qualitative ground for ascribing three-dimensionality to ordinary space. It is, however, clear that his proof demonstrates at best only the existence of a lower limit to the number of dimensions.

In an article written in the last year of his life Poincaré develops these ideas in detail. He says:

> The most important of all theorems of analysis situs is the statement that space has three dimensions . . . What do we mean when we say that space has three dimensions? . . . To separate space into parts, cuts are necessary which we call surfaces; to disconnect surfaces, cuts are necessary which we call lines; to divide lines, cuts are necessary which we call points. But we cannot go further, since a point, not being a continuum, cannot be divided. Therefore, lines that can be disconnected by cuts which themselves are not continua are continua of one dimension; surfaces that can be separated into parts by one-dimensional continua are continua of two dimensions; and finally space, which can be separated by two-dimensional continua, is a continuum of three dimensions.[118]

Poincaré was primarily interested in the physical and philosophical implications of the meaning of the concept of dimension, and yet this essay can be counted as the beginning of modern topological research concerning the mathematical problem of dimensionality. Although it was not his intent to formulate a strict mathematical definition of the concept of dimension, he anticipated the two essential elements of the modern definition of this term: the use of disconnecting subspaces and the inductive character of the definition. In fact, Brouwer's well-known topological invariant definition of dimension,[119] which for locally connected separable metric spaces is at present in common use, is based on Poincaré's considerations. Mathematicians, who until the beginning of the present century used the concept of dimension in a rather vague sense, became interested in a precise

[118] H. Poincaré, *Revue de métaphysique et de morale 20*, 486 (1912).
[119] L. E. J. Brouwer, "Ueber den natürlichen Dimensionsbegriff," *J. reine u. angew. Math. 142*, 146–152 (1913).

definition with the rise of the modern theory of sets. Cantor's famous one-to-one correspondence between the points of a line and the points of a plane, and Peano's continuous mapping of an interval on the whole of space, showed the deficiencies of the traditional definition of dimensionality as the smallest number of continuous real parameters sufficient to determine the position of a point. It was only in 1911 that Brouwer[120] established the proof that Euclidean spaces of different dimensionality are nonhomeomorphic, that is, cannot be mapped on each other by a continuous one-to-one correspondence. Important contributions by H. L. Lebesgue, K. Menger, P. Urysohn, and W. Hurewicz led to a further clarification of the mathematical concept of dimension.

To solve the problem of the dimensionality of space is also the ambition of modern physics. Among various attempts to this end the most noteworthy are probably those of Sir Arthur Eddington and H. Weyl. In his *Fundamental theory* Eddington succeeded by means of a complicated system of notions in reducing the problem to a reality investigation (in the mathematical sense of the word) of the so-called *E*-frame, a purely mathematical construct that is brought into physics to be identified with spacetime. Eddington says:

> The three-dimensionality of space and the time-like character of the fourth dimension are thus deduced directly from the properties of the *E*-frame. To what extent this amounts to an a-priori proof that the space-time of physical experience must be of this kind, depends on our inquiry into the ultimate origin of the *E*-frame in Chapter XIII.[121]

Unfortunately, Eddington died without having completed Chapter XIII. A note, probably written on the last day of his working life, indicates that the proposed chapter was to have been based

---

[120] L. E. J. Brouwer, "Beweis der Invarianz der Dimensionenzahl," *Math. Ann.* 70, 161–165 (1911).

[121] A. S. Eddington, *Fundamental theory* (Cambridge University Press, Cambridge, 1946), p. 124.

on his article on "The evaluation of the cosmical number." [122] Not enough material was published by him to furnish a decisive answer to the problem.

Weyl, in order to explain the three-dimensionality of space, refers to his generalization of Riemannian space to non-Riemannian gage-invariant geometry. He shows that only in a world conceived as a $(3 + 1)$-dimensional gage-invariant manifold (three spatial dimensions and one temporal dimension) does a most simple integral invariant exist in the form of action on which Maxwell's theory is founded.[123] The electromagnetic-field tensor is identified with "distance curvature" and Maxwell's equations appear as an intrinsic law.

And since it is impossible to construct an integral invariant at all of such simple structure in manifolds of more or less than four dimensions the new point of view does not only lead to a deeper understanding of Maxwell's theory but the fact that the world is four-dimensional, which has hitherto always been accepted as merely "accidental," becomes intelligible through it.[124]

The proof would be complete if it could be shown that all laws of gravitation as well as of electromagnetism are derivable from a variational principle that has to comply with the requirements of this invariance. It must, however, be admitted that Weyl's approach, ingenious as it is, is still open to serious objections.[125] In another attempt to include electromagnetic potentials in the metric of space, T. Kaluza[126] retained, in opposition to Weyl, the Riemannian character of space, but assumed an additional fifth dimension of the substratum underlying physical phenomena, thereby increasing the number of the components

[122] A. S. Eddington, *Proc. Camb. Phil. Soc. 40*, 37 (1944). See also Eddington, *Relativity theory of protons and electrons* (Cambridge University Press, Cambridge, 1936), chap. 6, "Reality conditions," and p. 325.
[123] H. Weyl, *Sitzber. preuss. Akad. Wiss.* (1918), p. 465; *Ann. Physik 59*, 101 (1919).
[124] H. Weyl, *Space-time-matter* (London, 1922), p. 284.
[125] See P. G. Bergmann, *Introduction to the theory of relativity* (Prentice-Hall, New York, 1950), p. 253.
[126] T. Kaluza, *Sitzber. preuss. Akad. Wiss.* (1921), p. 966.

of the metrical tensor. Similar procedures were employed in the so-called projective field theories of Veblen,[127] Hoffmann,[128] and Pauli,[129] and can be compared to the well-known representation in geometry of an n-dimensional space by $(n + 1)$ homogeneous coördinates. The use of five-dimensional tensors gained much popularity in other generalizations of general relativity, as for example in the modification introduced by Einstein and Mayer[130] and similar developments,[131] in particular in connection with the ambitious intention to account relativistically for the results of quantum mechanics. In conclusion of our treatment of the problem of dimensionality we may state that up to date no satisfying solution has been given. H. Grassmann's words, announced in 1844, have not yet been disproved:

> The concept of space can in no way be produced by thought, but always stands over against it as a given thing. He who tries to maintain the opposite must undertake the task of deducing the necessity of the three dimensions of space from the pure laws of thought, a task whose solution presents itself as impossible.[132]

Heisenberg, in his attempt[133] to achieve a simplified general representation of quantum mechanics, tried recently to abandon the principle of continuity in Riemannian or Euclidean geometry and introduced the suggestion of a "smallest length" to meet certain difficulties in quantum electrodynamics. This introduction of a discrete space with a quantum of length — Margenau[134] calls it a "hodon" from the Greek *hodos*, path, in analogy with the term "chronon" — would lead to a drastic revolution in the whole of theoretical physics. All differential equations would

[127] O. Veblen, *Projektive Relativitätstheorie* (Berlin, 1933).
[128] O. Veblen and B. Hoffmann, "Projective relativity," *Phys. Rev. 36*, 810–822 (1933).
[129] W. Pauli, *Ann. Physik. 18*, 337 (1933).
[130] A. Einstein and W. Mayer, *Sitzber. preuss. Akad. Wiss.* (1931), p. 541; (1932), p. 130.
[131] Cf. the concluding chapters in Bergmann's *Introduction to the theory of relativity*.
[132] Grassmann, *Die Ausdehnungslehre* (ed. 2, Leipzig, 1878), p. xxiii.
[133] Earliest reference, W. Heisenberg, Z. *Physik 110*, 251 (1938).
[134] Henry Margenau, *The nature of physical reality* (McGraw-Hill, New York, 1950), p. 155.

have to be recast into difference equations for the solution of which mathematicians would have to face almost insurmountable difficulties, although the subject of a finite geometry of discrete space structure has already been investigated, in particular by O. Veblen and W. H. Bussey.[135] From the historical point of view it is interesting to note that this possibility had also been envisaged already by Riemann. In the remarkable passage quoted on page 159 he said:

> If in a case of a discrete manifold the basis for its metrical determination is contained in the very idea of this manifold, then for a continuous one it should come from without. The reality which lies at the basis of space, therefore, either constitutes a discrete manifold, or the basis of metrical determination must be sought outside the manifold in the binding forces which act on it.

For the introduction of a "system of linkages" impressed upon a discrete manifold as a substitute for the fundamental tensor, the reader is referred to L. Silberstein's *The theory of relativity.*[136]

The concept of a *smallest length,* or rather a *fundamental length,* as characterizing the ultimate limit of resolution in physical measurement of spatial extension, has recently gained some popularity amongst theoretical physicists. Apart from Heisenberg, as mentioned above, A. March in particular advocated the assumption of a universal smallest length l_o.[137] A physical theory of spatial extension, he claims, has to be built on concepts that can be specified by their operational contents. The traditional geometry of points and infinitesimal magnitudes, therefore, has to be discarded as far as its immediate application in atomic physics is concerned. For any measurement is based ultimately on the coincidence of a scale with the object to be measured; and for the physicist the elementary particle is the smallest scale (or unit) available. The application of a concept of a still smaller spatial extension, not to say a pointlike extension, must — according to this school of thought — lead inevitably to insurmountable

[135] O. Veblen and W. H. Bussey, *Trans. Am. Math. Soc. 7,* 241 (1906).
[136] L. Silberstein, *The theory of relativity* (London, 1924), p. 362.
[137] Arthur March, *Natur und Erkenntnis* (Springer, Vienna, 1948).

difficulties. Indeed, the idea of a pointlike electron, for example, would imply the concentration of an infinite energy, while the conception of an extended rigid electron would contradict the principle of relativity. Since two particles whose distance apart is less than l_0 cannot be distinguished by diffraction experiments, l_0 becomes a universal length independent of the particular character of the particle in question.

The traditional conception of point coincidences has to be replaced by the notion of coincidences of particles. The spatial extension of elementary particles becomes apparent from the fact that a coincidence of particles A and B on the one hand, and of particles B and C on the other, does not necessarily lead to the result that A and C also coincide. The distance between two particles is determined by the minimal number of particles necessary to form a "chain of coincidences" between the given particles. Distances are therefore always integral multiples of l_0. The repeated occurrence in atomic physics of a length of the order of 10^{-13} cm, as the classical radius of the electron, the range of nuclear forces, or the critical energy of 10^8 electron volts, corresponding to a wave length of 10^{-13} cm, leads to the assumption that this length may be identified with l_0.

In view of the great mathematical difficulties involved in the construction of a geometry of discontinuous space, however, physics has still to resort to the traditional geometry of a continuous space by a statistical treatment of the concept of length. Thus continuous space resumes its service, even for nuclear physics, but as a convenient fiction for the statistical mathematization of physical reality.

A similar result seems to follow from even more general considerations. A profound epistemological analysis of certain quantum-mechanical principles seems to suggest that the traditional conceptions of space and time are perhaps not the most suitable frame for the description of microphysical processes. Thus Heisenberg's uncertainty principle states that the uncertainty involved in the measurement of the coördinate x of a

particle and the uncertainty involved in the simultaneous determination of the momentum p are governed by the relation
$\Delta x \cdot \Delta p \geqq h$ (h is Planck's constant). The impossibility of an
exact localization in combination with the determination of the
momentum, and the related dualistic wave-particle character of
physical reality, can be interpreted as a challenge for a critical
revision of the accepted space and time conceptions. In his discussion of electron transitions between stationary states within
the atom, Niels Bohr already called such processes "transcending
the frame of space and time." The problem of an intelligent
applicability of traditional space and time conceptions to atomic
physics was the subject of a paper submitted by Louis de Broglie
to the Tenth International Congress of Philosophy (Amsterdam,
August 11–18, 1948). De Broglie admits frankly the difficulties
involved in the use of our notions of space and time on a microphysical scale, but also confesses in the conclusion of his article
that up to the present no alternative conceptual categories are
known that can be substituted. He says:

> Les données de nos perceptions nous conduisent à construire un
> cadre de l'espace et du temps où toutes nos observations peuvent se
> localiser. Mais les progrès de la Physique quantique nous amènent à
> penser que notre cadre de l'espace et du temps n'est pas adéquat à
> la véritable description des réalités de l'échelle microscopique. Cepen-
> dant, nous ne pouvons guère penser autrement qu'en termes d'espace
> et de temps et toutes les images que nous pouvons évoquer s'y ratta-
> chent. De plus, tous les résultats de nos observations, même celles qui
> nous apportent le reflet des réalités du monde microphysique, s'expri-
> ment nécessairement dans le cadre de l'espace et du temps. C'est
> pourquoi nous cherchons tant bien que mal à nous représenter les
> réalités microphysiques (corpuscules ou système de corpuscules) dans
> ce cadre qui ne leur est pas adapté.[138]

Not only the notion of continuity, but also the conception
of "emptiness" have recently been subjected again to critical
examination. P. W. Bridgman, the keen operational analyst of
physical concepts, expounded the dilemma raised by submitting

[138] L. de Broglie, "L'espace et le temps dans la physique quantique,"
Proceedings of the tenth international congress of philosophy (North-Holland
Publishing Co., Amsterdam, 1949), vol. 1, p. 814.

the concept of empty space to the operational point of view.[139] Clearly no instrumental method can exist for testing such emptiness. The mere introduction of an instrument for this purpose already invalidates the very conditions of the situation under test. Moreover, no theory that attempts to eliminate the perturbations caused by the test body or test instrument (for example, a thermometer) can be applied to this situation, for such a theory has to be based on the variations in the reading of the instrument under changing conditions. Still "the intellectual compulsion remains to give some instrumental meaning to the purported emptiness of space. The simplest way of meeting this compulsion is simply to say that the space is empty if *no* instrument gives any reading when introduced into it." [140] But even this highly problematic specification seems to be untenable if confronted with the concept of an electrostatic field with fluctuating zero point, as advanced recently by quantum mechanics. If physics has to maintain the idea of empty space, it seems to be possible only by "ignoring part of the operational background." [141]

It will be recollected that it has been suggested by Riemann and Clifford, and later ingeniously corroborated by Einstein in his theory of general relativity, that the metric of space structure is a function of the distribution of matter and energy. In accordance with this principle the large-scale properties of space became the object of cosmological research during the last three or four decades. In fact, Einstein's paper on "Cosmological considerations in general relativity," [142] published in 1917, turned the investigations of macroscopic space structure (for example, the question whether space is finite or infinite) and cosmology in general from poetical and philosophical speculations into a

[139] P. W. Bridgman, *The nature of some of our physical concepts* (Philosophical Library, New York, 1952).
[140] *Ibid.*, p. 19.
[141] *Ibid.*
[142] A. Einstein, "Kosmologische Betrachtungen zur allgemeinen Relativitätstheorie," *Sitzber. preuss. Akad. Wiss.* (1917), p. 142.

respectable scientific discipline with solid foundations in physics, in spite of the fact that many of its most important issues are still under debate. The article has shown incontestably that the mathematical apparatus of general relativity, when applied to cosmological problems, could bring science a decisive step nearer to a solution of the large-scale problems of space.

It has been shown by H. P. Robertson[143] and independently by A. G. Walker[144] that essentially only the following metric is compatible with the assumption of a large-scale homogeneous isotropic space-time continuum:

$$ds^2 = dt^2 - \frac{[R(t)]^2\,[(dx^1)^2 + (dx^2)^2 + (dx^3)^2]}{\{1 + \tfrac{1}{4}k[(x^1)^2 + (x^2)^2 + (x^3)^2]\}^2},$$

where k may have the values 0, -1, or 1. Confining our consideration to purely spatial extension [$t = $ const. and $R = R(t)$], we see that if $k = 0$, space is Euclidean: the surface of a sphere with radius r is $4\pi r^2$. If $k = -1$, space is hyperbolic: the surface of a sphere with radius r is $64\pi R^2 \sinh^2 (r/4R)$, that is, greater than $4\pi r^2$, as can be seen at once from the expansion into a power series. Finally, if $k = +1$, space is spherical: the surface of a sphere with radius r is $64\pi R^2 \sin^2 (r/4R)$, that is, less than $4\pi r^2$. We see from the last formula that a sphere has a maximum surface if its radius is equal to $2\pi R$; if its radius is greater than twice the "radius of the universe R," the surface of the sphere decreases until for $r = 4\pi R$ it shrinks to a point. A null-geodesic forms a closed curve, that is, light rays return to their starting point. If the Robertson line element is transformed to polar coördinates, a simple integration shows that Euclidean and hyperbolic spaces are unlimited in volume, whereas spherical space, at any instant t, has a total finite volume $2\pi^2 R^3$.

It is well known that Einstein's classical paper of 1917 characterized space as endowed with the last-mentioned of these properties. The inference of a finite total volume of space was a

[143] H. P. Robertson, *Proc. Nat. Acad. Sci. 15,* 822 (1929).
[144] A. G. Walker, *Proc. London Math. Soc. 42,* 90 (1936).

consequence of the postulated isotropy and homogeneity of space together wtih the assumption of a constant finite density ρ_o of the stellar masses at rest. (Only fifteen years later did Edwin Hubble discover the red shift of the spectral lines emitted by nebulae.) Since these three assumptions proved to be incompatible with the original field equations of 1915,[145] Einstein proposed in 1917 [146] to make the theory consistent by modifying the field equations and thus introduced the cosmological constant λ. For only the introduction of a suitable constant could make all components of the energy-momentum tensor $T_{\mu\nu}$ equal to zero with the exception of the component T_{oo} which represents the density of matter in the world. The modified equations admit of a solution consistent with the three above-mentioned assumptions, namely that it describes a space of finite volume the radius of which is related to the average density of matter ρ_o through the field equations.

Apart from reconciling the theory with these assumptions, the introduction of the cosmological constant had to satisfy another need which in Einstein's view was of great significance for the philosophical conception of space: it had to implement Mach's Principle within the conceptual frame of general relativity or at least to remove certain difficulties that prevented an incorporation of the Principle into the general theory of relativity. Since the status of Mach's Program in its relation to general relativity is intimately connected with the problem of absolute space a more detailed statement of the pertinent results seems to be appropriate.

The classical law of inertia states that a physical body, once released in its motion in a given world direction (initial conditions), in the absence of external forces pursues an invariable course mapped out by the guiding field. The amalgamation of inertia and gravitation which according to Einstein determines

[145] A. Einstein, Feldgleichungen der Gravitation, *Sitzber. preuss. Akad. Wiss.* (1915), pp. 844–847.
[146] See reference 142.

the guiding field and replaces the purely Galilean inertial tendency by an infinitesimal parallel displacement along a geodesic line, raises the question of whether this guiding field is exclusively determined by the large-scale distribution of matter and its relative motion. In other words: is the guiding field (the world metric) of general relativity exhaustively determinable by the energy-momentum tensor alone?

The statement that the kinematic-dynamic behavior of bodies is determined by the distribution of matter in the universe is generally referred to as Mach's Hypothesis. In a note to the second edition of his *Science of Mechanics* (*Die Mechanik in ihrer Entwicklung*),[147] Mach raises the following problem: does a fourth mass-point *D* (cf. page 139 of the present book), when left to itself and when not subjected to any forces, describe a straight line with constant velocity relative to the "inertial system *S*" as defined by the projection of the other three "free" mass-points *A*, *B* and *C* in the absence of the fixed stars or in the case of large-scale changes amongst them? In later editions of his *Science of Mechanics*, Mach generalized this idea and proclaimed as his program the establishment of a principle on the basis of which "inertial and accelerated motions could equally be accounted for." (*"Eine prinzipielle Einsicht, aus der sich in gleicher Weise die beschleunigte und die Traegheitsbewegung ergeben . . ."*) Mach's Principle, as originally announced, claimed the intrinsic dependence of every local inertial system, that is, a local coordinate system in which Newton's laws hold, upon the distribution of mass in the universe.

Mach's juxtaposition of inertial and accelerated (gravitational) motion may be interpreted as though gravitational motion in Newtonian dynamics is subject to such a principle. Strictly speaking, Newtonian gravitational theory does not comply with Mach's Program since the determination of the potential is only possible

---

[147] See reference 35. The first edition of Mach's *Mechanik* appeared in 1883 (Brockhaus, Leipzig), the second in 1889. The above mentioned note is found in the second edition on p. 485.

if, in addition to the Laplace-Poisson differential equation, certain boundary conditions at infinity are supplied (as, e.g., that
the potential vanishes at infinity). The inclusion of such boundary conditions, however, violates Mach's Principle.

The original conception of the principle, as expressed by Mach,
based on the assumption of instantaneous interaction, is of course
without further modifications (advanced potentials, etc.) inad-
missible for the general theory of relativity. In a short paper
entitled *Prinzipielles zur Allgemeinen Relativitaetstheorie* ("Principles concerning the General Theory of Relativity")[148] Einstein
discusses the leading ideas of his paper of 1917 and in particular
his attempt at a synthesis of Riemannian geometry with Mach's
Principle. With reference to the proposed ontological subordination of space-time to matter, Einstein says: "I have chosen the
name Mach's Principle because the Principle implies a generalization of Mach's requirement according to which inertia should
be reduced to the interaction of bodies." Einstein deems the
implementation of this Principle as "absolutely necessary" (*"als
unbedingt notwendig"*), although he admits that his opinion in
this matter is not shared by all the scientists of the time. If indeed
the structure of space-time is exhaustively conditioned and determined by the distribution of masses, he claims, the absence of
matter should imply the nonexistence of a guiding field (space-
time metric). The original (unsupplemented) field equations
(see page 159) obviously do not satisfy the Principle. Einstein's
introduction of the cosmological constant λ, by which he hoped
to remove the inconsistency with Mach's Principle, stands in
striking similarity to H. Seeliger's modification of the classical
Laplace-Poisson equation $\Delta\varphi = 4\pi\rho$ into $\Delta\varphi - \mu^2\rho = 4\pi\rho$, whereby
Seeliger attempted to relieve Newtonian cosmology from certain
inconsistencies.[149] The positive constant μ should be chosen so

[148] *Ann. d. Physik,* 55, 241–244 (1918).

[149] H. Seeliger, Über das Newton'sche Gravitationsgesetz, *Astr. Nachr.*
137, 129–136 (1895). Cf. also *Sitzber. Münchener Akad. Wiss.* 26, 373–400
(1896). For further details cf. also M. Jammer, *Concepts of Mass,* (forthcoming), chapter 10.

small that within the dimensions of the solar system the solution of the original equation (i.e., $\varphi = -m/r$) and that of the supplemented equation $(\varphi = -\dfrac{m}{r}e^{-\mu r})$ should coincide within the margin of observational error.

Similarly, Einstein expected that the modified field equations

$$R_{\mu\nu} - \lambda g_{\mu\nu} = -k\,T_{\mu\nu} - \tfrac{1}{2}g_{\mu\nu}T$$

would implement Mach's Principle within the framework of the general theory of relativity without affecting the observational results pertaining to not too large distances. Einstein declared: "A space-time continuum, free of singularities, with everywhere vanishing energy-momentum tensor seems to be precluded by the modified field equations."

Einstein's statement, however, was soon refuted by W. de Sitter [150] who showed that the *ad hoc* modified field equations admit of a space-time structure as a solution even in the absence of matter. The assumption of a completely empty universe, containing neither matter nor radiation, but endowed with space-time structure and consequently with a guiding field (and inertia) is obviously incompatible with Mach's Principle.

On the other hand, various solutions of dynamical problems on the basis of Einstein's field equations seemed to speak in favor of Mach's Principle. Thus Hans Thirring, for example, in a paper *On the effect of distant rotating masses in Einstein's theory of gravitation* [151] showed that in the gravitational field of general relativity distant rotating masses subject a test-body to accelerations which are perfectly analogous to those produced by the centrifugal or Coriolis forces of classical dynamics. In the case of a rotating spherical shell the mathematical expression for these forces contains a factor $(1 + 2kM/a)$ where a and M denote the radius and the mass, respectively, of the shell. The fact that this

[150] On Einstein's Theory of Gravitation and its Astronomical Consequences, parts 1, 2, 3. *Monthly Notices R.A.S.*, 76, 699 (1916), 77, 155 (1916), 78, 1–28 (1917).
[151] Über die Wirkung rotierender ferner Massen in der Einsteinschen Gravitationstheorie, *Phys. Z.*, 19, 33–39 (1918), 22, 29–30 (1921).

factor depends on the mass and on the radius of the rotating shell (the source of the field of forces) seems to confirm the validity of Mach's Principle. It can also be shown that for static fields with a metric $ds^2 = da^2 - v^2/c^2(dx^4)^2$, where $da^2 = g_{ik}dx^idx^k$ ($i, k = 1, 2, 3$) and where v represents the time-independent but locally variable velocity of light, the following equations of motion are first-order solutions: [152]

$$\frac{d}{dt}\left(\frac{cm}{v}\frac{dx^i}{dt}\right) = v\,K^i - m\,c\frac{\partial V}{\partial x^i}\,(i = 1, 2, 3).$$

K^i is the contravariant component of the Minkowski force, and the partial derivative on the right-hand side is an expression for the gravitational force, analogous to the gradient of the Newtonian gravitational potential. Inertial mass is apparently represented not by m but by $\left(\frac{c}{v}m\right)$, an expression which through the variability of v (as a function of the gravitational potential) depends on the distance from the gravitational masses. On the surface of the sun, for instance, the inertial mass would increase by two millionths of its interstellar value, a result which again seems to plead the cause of Mach's Principle. A more rigorous derivation of the equations of motion, however, will show that the inertial mass of a body in a gravitational field is independent of its location.

Nor can the examples referred to, even without considering the last-mentioned objection, be construed as a demonstration of the validity of Mach's Principle, since, partially under disguise of the method of approximation, the Minkowskian or pseudo-Euclidean space-time structure was tacitly assumed to hold at infinity. Just as Newtonian dynamics in solving Poisson's equation had to have recourse to boundary conditions at infinity, so in the present problems, the solution of the field equations presupposes similar boundary conditions at infinity. But to assume that the components of the fundamental metric tensor $g_{\mu\nu}$ have the

[152] Cf. e.g. Max von Laue, *Die Relativitaetstheorie*, vol. 2 (Vieweg, Braunschweig, 1953), p. 123.

Lorentz-Minkowski values at infinity violates Mach's Principle in
the following twofold sense, as A. Grünbaum [153] has pointed out:
"(a) the boundary conditions at infinity assume the role of Newton's absolute space, since it is *not* the influence of matter that
determines what coordinate systems at infinity are the Galilean
ones of special relativity; and (b) instead of being the *source* of
the *total* structure of space-time, matter then merely *modifies*
the latter's otherwise autonomously flat structure."

Recently Mach's Program seemed to have lost further ground
when H. A. Taub showed for the unmodified field equations that
under certain conditions (space-times admitting a three parameter continuous group of motions described by spacelike
infinitesimal generators) the flatness of space-time, *i.e.*, the vanishing of the Riemann-Christoffel tensor $R^\mu_{\nu\sigma\tau}$, is a necessary consequence of the absence of matter ($T_{\mu\nu} = 0$) *only provided $R_{\mu\nu\delta\tau}$*
is not singular along the "time" axis (the "time" dependence of
the $g_{\mu\nu}$ is determined under these conditions by the field
equations). Otherwise, the coefficients of the fundamental metric
possess "spatial singularities," that is, in a special coordinate
system they are not bounded for all values of the spatial coordinates. If these singularities are attributable to the coordinate
system alone, they are not "essential" in the sense that they represent matter. Consequently, absence of matter and space-time
curvature are not necessarily mutually exclusive conceptions.[154]

These results seem to indicate that Mach's Principle is neither
a logical consequence of the general theory of relativity nor a
necessary presupposition of the same. Its compatibility with the
theory, however, is still a matter of dispute.[155] Should it become

[153] Adolf Grünbaum, The Philosophical Retention of Absolute Space in
Einstein's General Theory of Relativity, *The Philosophical Review 66*, 525–
534 (1957).
[154] A. H. Taub, Empty Space-Times admitting a three parameter group of
motions, *Ann. Math. 53*, 472–490 (1951).
[155] Cf. in this respect F. A. Kaempffer, On possible realizations of Mach's
Program, *Can. J. Phys. 36*, 151–159 (1958). A tentative theory to account
for inertia in strict conformity with Mach's Principle has been advanced by
D. W. Sciama, *Monthly Notices R.A.S. 113*, 34–42 (1953).

clear that Mach's Principle is compatible with the general theory
of relativity, matter and space-time could be considered as two
ultimately distinct physical entities and the Newtonian notion of
absolute space would have to be eliminated from the conceptual
scheme of theoretical physics. Should it, however, become evident
that Mach's Program cannot be satisfied within the general
theory of relativity perhaps merely because the energy-momentum
tensor $T_{\mu\nu}$ which characterizes matter presupposes already metrical magnitudes: in other words, because matter cannot be understood apart from knowledge of space-time, then matter as the
source of the field will become part of the field. On the basis of
such a unified field theoretic conception as proposed for example
by J. Callaway,[156] the field itself would constitute the ultimate,
and in this sense absolute, datum of physical reality. As the concluding words in his Foreword to the present book seem to imply,
Einstein's view during the last years of his life was in favor of this
conception. Although matter may provide the epistemological
basis for the metrical field, it does not necessarily have ontological
priority over the field.

Recent studies of this problem and in particular the work by
Ozsváth and Schücking[157] seem to lend support to the conclusion that general relativity fails to exclude solutions which contradict Mach's Principle. The results obtained so far thus seem to
indicate that theoretical physics — at least to the extent as it exists
in full-fledged form at present — has not yet succeeded in exhaustively subordinating space to matter, and Newton's ghost of absolute space has not yet been completely exorcised. In view of these
facts, there exist at present two major schools of thought concerning Mach's Principle and its impact upon the concept of space.

[156] J. Callaway, Mach's Principle and Unified Field Theory, *Phys. Rev. 96*,
778–780 (1954).
[157] I. Ozsváth and E. Schücking, "Finite rotating universes," *Nature 193*,
1168–1169 (1962). But cf. also H. Dehnen and H. Hönl, "Finite universes and
Mach's principle," *Nature 196*, 362–363 (1962).

One is the Princeton School, led by Dicke[158] and his pupil Brans,[159] which tries to modify the general theory of relativity to such an extent that by treating gravitational effects in terms of a scalar-tensor field in a Riemannian manifold, Mach's Principle is fully incorporated into the theory. The other is the Freiburg School, led by Hönl,[160] which regards Mach's thesis as a principle of selection for cosmological models. Since all finite models, such as Einstein's static universe, seem to satisfy Mach's Principle, Hönl regards it — and this has a great philosophical importance — as merely a relaxed form of a general postulate which demands the exclusion of actual infinities. The very existence of opposing schools of thought clearly indicates that a final solution of the problem has not yet been reached.

Another question whose answer is not yet in sight and which has engaged the interest of many scientists during the past few years is the problem of parity. It brings us back to Kant's famous "proof" for the existence of absolute space on the basis of the distinction between right and left.[161] In order to understand fully the recent developments on this question, let us reformulate Kant's argument as follows: (1) there are enantiomorphic objects A and A', like right and left hands, which cannot be put into coincidence although they possess the same intrinsic geometry (position relations of parts); (2) if God's first creative act had been the forming of a left hand, this hand, even if it could not yet be compared to anything else, would have possessed "left-handedness"; (3) hence, there exists a spatial property p which is possessed by A but not by A'; (4) hence, there are spatial properties of objects not reducible to the position of their parts; (5) hence, the property p must be

[158] R. H. Dicke, "Mach's principle and equivalence," in *Evidence for gravitational Theories* (Academic Press, New York, 1962).

[159] C. Brans and R. H. Dicke, "Mach's principle and a relativistic theory of gravitation," *Physical review 124*, 925–935 (1961).

[160] H. Hönl and H. Dehnen, "Erwiderung auf die Arbeit von J. Ehlers and E. Schücking über die Formulierung des Machschen Prinzips," *Zeitschrift für Physik 206*, 492–502 (1967).

[161] See p. 133.

grounded in a relation of A to space; (6) but space, endowing objects with real properties, must have reality itself.

Recently, Kant's argumentation was challenged by Reidemeister[162] on the grounds that a distinguishability between A and A' does not necessarily imply the existence of a property p which is exhibited by A but not by A' or *vice versa*. Relations *between* objects, Reidemeister argued, need not be reducible to relations that are individually inherent in the objects themselves. In support of his theory Reidemeister showed in great detail how the whole of Euclidean geometry — with the inclusion of the notion of oppositely oriented lines etc. — can be axiomatized exclusively in terms of distance between points which *qua* points have no intrinsic properties *per definitionem*. Reidemeister's refutation of Kant was criticised by Lange,[163] whose conception of Kant's philosophy of space differs fundamentally from that of Reidemeister but shows some similarity to that of Kurth[164] in spite of the fact that Lange referred primarily to the pre-critical and Kurth to the critical stage of Kant's philosophy. Kant's demonstration has also been challenged recently by Remnant,[165] who even calls into question the correctness of Kant's premise concerning the identity of the *intrinsic* geometry of enantiomorphic objects. Although Remnant refers repeatedly to the two-dimensional analogue of two similar but incongruent scalene triangles of equal area which by rotation through the third dimension can readily be brought into coincidence, he seems to have failed to realize that two three-dimensional enantiomorphic objects can also be brought into coincidence if rotated in a four-dimensional space. In fact, that two n-dimensional objects, symmetric with respect to an $(n-1)$-dimensional object, can be

[162] K. Reidemeister, *Raum und Zahl* (Springer, Berlin, Göttingen, Heidelberg, 1957), pp. 53–69.

[163] H. Lange, "Über den Unterschied der Gegenden im Raume," *Kantstudien 50*, 479–499 (1958–1959).

[164] R. Kurth, "Kant's Lehre von Raum und Zeit," *Philosophia naturalis 4*, 266–296 (1957).

[165] P. Remnant, "Incongruent counterparts and absolute space," *Mind 72*, 393–399 (1963).

brought into coincidence by a rotation in $(n + 1)$-dimensional space had been proved already in the last century.[166]

If the preceding considerations touched upon the problem of parity from the logico-geometrical point of view, a startling experimental discovery in 1956 turned it into a problem of far-reaching physical significance.

The dynamical behavior of a physical system is generally described by integral or differential equations in which the unknowns are either certain functions (as in classical dynamics) or proper values of certain operators (as in quantum mechanics). In all physically important cases it is possible to derive these equations from a variational principle, as e.g., that of least action. A conservation law, such as the conservation of momentum, of angular momentum, or of energy, then does not actually appear as a new law of nature but is merely the mathematical expression of an invariance that originates in certain symmetry properties of the respective action function or action functional on the basis of which the equations of motion have been formulated.

The relation between symmetry properties and conservation laws, at least from the mathematical point of view, had already been fully explained on the basis of Lie's theory of continuous groups in 1918 by Emmy Noether.[167] As an elementary example, consider the Hamiltonian representation of a dynamically closed system with the azimutal angle φ as an ignorable coordinate. Owing to the isotropy of space the Hamiltonian function H is independent of φ. According to the well-known Hamilton equation $\dot{p} = -\partial H / \partial q$, applied for the coordinate $q = \varphi$, the conjugate momentum p, in this case the angular momentum, is a constant of motion (i.e., independent of time). Thus, the conservation of angular momentum is a consequence of the isotropy or rotational symmetry of space. Likewise, the conservation of linear

---

[166] Cf. J. Delboeuf, "L'ancienne et les nouvelles géometries (2ème)," *Revue philosophique de la France et de l'étranger 37*, 353–383 (1894).

[167] Emmy Noether, Invariante Variationsprobleme, *Goettinger Nachr.*, *1918*, 235–257.

momentum is a consequence of the translational symmetry or homogeneity of space.

These conservation laws were not the only ones that modern physics had to cope with. In 1924, Otto Laporte[168] in connection with his investigation of the structure of the iron spectrum called attention to certain regularities among the spectral lines and their intensities. Three years later E. Wigner[169] explained these regularities as a consequence of the conservation of parity, that is, invariance under inversion of coordinates: $x \longrightarrow -x$; $y \longrightarrow -y$; $z \longrightarrow -z$. It was immediately clear, and in fact it can be demonstrated in an elementary manner,[170] that classical mechanics as well as classical electrodynamics satisfy the principle of parity conservation. It was tacitly assumed that the parity conservation principle holds not only for the universe of macroscopic bodies but also for the microscopic world of elementary particles. A preference for a given handedness or in other words an intrinsic asymmetry of space seemed inconceivable.

In 1956, however, certain experimental results concerning the modes of decay of charged K-particles made T. D. Lee and C. N. Yang suspect that the parity conservation law may possibly not hold for weak interactions. In their provocative paper entitled *Question of Parity Conservation in Weak Interactions*[171] they declared: "It will become clear that existing experiments do indicate parity conservation in strong and electromagnetic interactions to a high degree of accuracy, but that for weak interactions (*i.e.*, decay interactions for the mesons and hyperons, and various Fermi interactions) parity conservation is so far only an extrapolated hypothesis unsupported by experimental evidence. . . . To decide unequivocally whether parity is conserved in weak inter-

[168] Otto Laporte, Die Struktur des Eisenspektrums, Z *f. Phys.*, *23*, 135–175 (1924).

[169] E. Wigner, Einige Folgerungen aus der Schroedingerschen Theorie für die Termstrukturen, *ibid.*, *43*, 624–652 (1927).

[170] Cf. David Park, Recent Advances in Physics, *Am. J. Phys.*, *26*, 210–234 (1958) or Ta-You Wu, Laws of Conservation: Parity and Time Reversal, *ibid.*, 568–576.

[171] *Phys. Rev.*, *104*, 254–258 (1956).

actions one must perform an experiment to determine whether weak interactions differentiate the right from the left. Some such possible experiments will be discussed."

In order to force a decision in this question whether parity conservation applies also to weak interactions, the probability distribution of a pseudoscalar (*i.e.*, one which unlike ordinary scalars changes sign under inversion, such as the scalar triple product of polar vectors $a \cdot (b \times c)$ or the projection of a polar vector along an axial vector, $p \cdot I$, where p is a polar vector and I an axial vector) had to be measured.

The experiments suggested by Lee and Yang were performed by C. S. Wu, E. Ambler, R. W. Hayward, D. D. Hoppes and R. P. Hudson[172] who succeeded in aligning Co^{60} nuclei (possessing a large magnetic moment due to their spin of 5) by applying strong magnetic fields at temperatures below $1°$ Kelvin. The β-active nuclei were thus polarized in the sense that all their intrinsic angular momenta I (axial vectors) pointed in the same direction. The momentum p of the emitted electrons is a polar vector and the scalar product $p \cdot I$, as explained above, a pseudoscalar. The angular distribution (angle between p and I) of the emitted electrons relative to the spin of the mesons was measured and a marked asymmetry was observed.

It was startling news. It was as if suddenly an experiment revealed that linear momentum (in a closed dynamical system) is not preserved and space is no longer to be conceived as homogeneous. The only difference was that experimental evidence referred to rotational properties and not to translational ones. The correspondence between W. Pauli and V. Weisskopf[173] shows eloquently to what extent even competent experts were taken by surprise.

[172] Experimental Test of Parity Conservation in Beta-Decay, *Phys. Rev.* *105*, 1413–1415 (1957).

[173] As Professor Weisskopf informed the present author this correspondence is to be published shortly. Cf. Abdus Salam, Elementary Particles and Space-Time Symmetry, *Endeavour 17*, 97–105 (1958).

Later, Garwin and collaborators,[174] Friedman and Telegdi[175] and others provided additional experimental evidence for the nonconservation of parity. On the basis of these and similar results Lee and Yang,[176] Salam[177] and Landau[178] advanced the "two-component theory" in an attempt to explain the apparent violation of parity conservation as a consequence of a screw-asymmetry of the neutrino and the antineutrino which are known to be emitted simultaneously with the positron and negatron in the β-decay process. The spin of the neutrino is always right-handed with respect to the direction of propagation while that of the antineutrino is always left-handed. The mirror copy of the neutrino (or of the antineutrino, respectively) has no physical reality.

The particle-antiparticle relation presented in this process, the fact that the antineutrino is always emitted in combination with the electron and the neutrino with the positron, seems to suggest that the β-decay of antimatter is characterized by a screw-asymmetry which is inverse to that of matter. L. Landau who in this connection defined the concept of "combined inversion" as the simultaneous occurrence of spatial inversion and charge conjugation regards the simple rejection of parity conservation as untenable for theoretical physics. He says: "It is easy to see that invariance of the interaction with respect to combined inversion leaves space completely symmetrical, and only the electrical charges will be asymmetrical. The effect of this asymmetry on the symmetry of space is no greater than that due to chemical stereoisomerism." [179]

[174] R. L. Garwin, L. M. Lederman, M. Weinrich, Observations of the Failure of Conservation of Parity and Charge Conjugation in Meson Decays, *Phys. Rev. 105*, 1415–1417 (1957).
[175] J. I. Friedman and V. L. Telegdi, Nuclear Emulsion Evidence for Parity Nonconservation, *ibid.*, 1681–1682.
[176] Parity Nonconservation and a Two-Component Theory of the Neutrino, *ibid.*, 1671–1675.
[177] A. Salam, On Parity Conservation and Neutrino Mass, *Nuovo Cimento 5*, 299 (1957).
[178] L. Landau, On the Conservation Laws of Weak Interactions, *Nuclear Physics 3*, 127–131 (1957).
[179] *Loc. cit.*, 128.

In recent years the problem of the dimensionality of space, and in particular the question of how far contemporary physics implies necessarily a space of three dimensions or a space-time of four, gained renewed prominence. Some of these studies are elaborations of Ehrenfest's arguments of 1917 and 1920, which attracted little attention at that time.[180] Using a generalization of the Newtonian law of universal gravitation,

$$G M m / r^{n-1},$$

for the n-dimensional Euclidean space R_n, Ehrenfest showed that for $n > 3$ a perturbation, however small, would cause the planet (of mass m) either to spiral into the central body (of mass M) or to move into infinity; only if $n = 2$ or $n = 3$ is a perturbation compatible with the stability of the orbit.[181] Ehrenfest studied also the consequences of $n \neq 3$ for the Bohr model of the atom and found that a generalization of Coulomb's law, analogous to the preceding generalization of Newton's law, would lead (under preservation of the quantization of the angular momentum) to extreme degeneracies and other severe anomalies of the energy levels in contrast to spectroscopic experience. Referring to Hadamard's investigations of the Huygens' principle for wave propagation, Ehrenfest pointed out that electromagnetic signals could be transmitted in an undistorted manner only in odd-dimensional spaces.[182]

Ehrenfest's invocation of the Newtonian law of gravitation found a philosophically interesting elaboration in an argument

[180] P. Ehrenfest, "In what way does it become manifest in the fundamental laws of physics that space has three dimensions?" *Proceedings of the Amsterdam academy 20*, 200–209 (1917); "Welche Rolle spielt die Dreidimensionalität des Raumes in den Grundgesetzen der Physik?" *Annalen der Physik 61*, 440–446 (1920).

[181] Cf. J. Bertrand, "Théorème relatif au mouvement d'un point attiré vers un centre fixe," *Comptes rendus 77*, 849–853 (1873). See also H. Lamb, *Dynamics* (Cambridge University Press, Cambridge, Eng., 1914), pp. 256–258, where it is proved that in the case of a central force, proportional to r^{-n}, circular orbits are stable only if $n < 3$.

[182] For a nontechnical outline of these considerations, cf. K. Schäfer, "Die Zeit und die übrigen Dimensionen," *Studium generale 20*, 1–9 (1967).

advanced by Whitrow.[183] Starting from the fact that we human
beings *do* study this problem, Whitrow contends that we could not
have reached such an advanced stage in the process of biological
evolution had it not been for the favorable conditions on the earth's
surface of a more or less constant intensity of solar radiation
throughout hundreds of millions of years and thus for the stability
of a nearly circular orbit of the earth in the gravitational field of the
sun. Referring to the theorem demonstrated in Lamb's textbook of
dynamics,[184] Whitrow concludes that $n = 2$ or $n = 3$. To eliminate the former alternative, Whitrow refers to the existence of
typical multicellular animals with alimentary canals whose torus-
shaped bodies exhibit a topology incompatible with $n = 2$.

That there exist intimate and yet unexpected relations between
the theoretical foundations of electromagnetism and the three-
dimensionality of space — or rather the four-dimensionality of
space-time — and that apart from Maxwell's (relativistic) equations, other field-theoretic results, such as Dirac's equations of the
electron, manifest remarkable symmetries, not presentable in
spaces of another number of dimensions, has been recently discovered by Lanczos.[185] Furthermore, since all of our present laws
in physics admit, at least formally, of extensions to spaces with an
arbitrary number of dimensions, it may be suggested that there
exists some far-reaching principle which, in conjunction with these
laws, implies the empirical particularity $n = 3$ for spatial dimensionality. Generalizing the considerations advanced by Ehrenfest
and Whitrow with respect to the Newtonian-Keplerian problem in

[183] G. J. Whitrow, "Why physical space has three dimensions?" *The British
journal for the philosophy of science 6*, 13–31 (1955). Cf. also W. Büchler,
"Warum hat unser Raum gerade drei Dimensionen?" *Physikalische Blätter 19*,
547–549 (1963); *Philosophische Probleme der Physik* (Herder Verlag, Freiburg, Basel, Wien, 1965), pp. 151–156, where similar physiological arguments
are adduced to prove that $n \geq 3$ (topological restrictions in spaces with $n < 3$
would preclude the evolution of sufficiently developed nervous systems) and
where Ehrenfest's conclusions of $n \leq 3$ from the stability of the Bohr model of
the atom are rederived on the basis of the Heisenberg uncertainty relations.

[184] See note 181 above.

[185] C. Lanczos, "The splitting of the Riemann tensor," *Reviews of modern
physics 34*, 379–389 (1962).

n dimensions, Tangherlini[186] proposed a tentative formulation of such a principle by postulating that there exist stable bound orbits or "states" for bodies, treated as material points and interacting through fields which at large distances approach asymptotically to constant values. By applying this postulate to the geodesic equations of motion obtained by generalizing the Schwarzschild field to static systems with hyperspherical symmetry, Tangherlini succeeded in proving that his so-called "bound state postulate" does in fact entail uniquely the spatial dimensionality. He also studied the Schrödinger hydrogen atom in n dimensions and showed that the postulate also excludes $n > 3$ and, in conjunction with an asymptotic condition, $n < 3$ as well. More recently still, additional support to this trend of argumentation may be found in Penney's elaboration of Einstein's remark that the gravitational equations for empty space determine the metric just as "strongly" as Maxwell's equations do for the Maxwell field, the "strength" of a set of field equations being determined by the number of the kth-order Taylor coefficients in the expansion of the field function remaining free, that is not eliminable by taking account of the field equations. On the basis of this definition Penney[187] demonstrated that only in a space-time continuum of four dimensions do the equations of the gravitational field, of the electromagnetic field, and of the neutrino field, as described by the Weyl equation, determine their fields with equal strength. Since an inequality of strength entails a discontinuity of the metric, a physically unacceptable conclusion, it follows that the four-dimensionality of space-time is a necessary condition for the existence of a unified field theory embracing the three principal massless, long-range particles of Einstein, Maxwell, and Weyl.

Reichenbach's doctrine of the relativity of geometry[188] — that

[186] F. R. Tangherlini, "Schwarzschild field in n dimensions and the dimensionality of space problem," *Nuovo cimento 27*, 636–651 (1963).

[187] R. Penney, "On the dimensionality of the real world," *Journal of mathematical physics 6*, 1607–1611 (1965).

[188] H. Reichenbach, *The philosophy of space and time* (Dover Publications, New York, 1958), pp. 30–37.

is, his contention that the properties of physical space are objectively determinable by measurements with rigid rods only if combined with an accepted coordinative definition of congruence, and his conclusion that different geometries, such as Euclidean or non-Euclidean, can be ascribed to one and the same physical space depending on what coordinative definition has been chosen — was always referred by Reichenbach and his school to the metrical structure of space alone. In fact, the *relativity of geometry* has been regarded as an expression for the legitimacy of any remetrizability of physical space compatible with its topological properties. Among the latter the conservation of dimensionality was always considered a restrictive condition. That the relativity of geometry can be extended also to the number of dimensions, at least to a certain extent, was contended in 1965 by Stenius.[189] Using Peano's Curve, Stenius mapped by an essentially one-to-one continuous mapping (that is up to sets of zero Lebesgue measure) the entire one-dimensional Euclidean space on a two-dimensional Euclidean space and generalized this procedure for arbitrary dimensionalities. Whereas in the well-known remetrization processes from non-Euclidean to Euclidean geometry (Klein, Beltrami) rigid rods contract or expand and thus undergo continuous transformations, the behavior of solid bodies under the transformations discussed by Stenius is length-preserving but exhibits strange microphysical discontinuities. Stenius did not elaborate on the philosophically interesting, though highly speculative, problem of how these may be related to the discontinuities actually encountered in microphysics.

Apart from these questions pertaining to the topological properties of space, the problems related to the metrical properties of physical space also re-engaged the attention of contemporary philosophers and scientists, because of the work of Grünbaum.[190]

[189] E. Stenius, "On the number of dimensions of physical space," *Acta philosophica fennica 18*, 227–241 (1965).

[190] A. Grünbaum, "Conventionalism in geometry," in *The axiomatic method with special reference to geometry and physics* (North-Holland Publishing Company, Amsterdam, 1959), pp. 204–222; "Geometry, chronometry, and

Grünbaum's interest in the problem, in what sense and to what extent the choice of a particular metric can be held to have an empirical warrant, had been prompted by a conflict of opinion on this matter among Robertson, Reichenbach and Einstein.[191] By a penetrating analysis of the roles of convention and fact in the ascription of a particular metrical structure to physical space as a result of measurements performed with rigid bodies, Grünbaum traced the core of the problem to the epistemological status of congruence. This was for Newton, as we have seen, an intrinsic relation of space, for Riemann a function of the coordinates, expressed by the components of the metric tensor, and for Poincaré a matter of convention. Grünbaum critically compared these conceptions of congruence as well as those advanced by Eddington, Bridgman, Russell, and Whitehead. And he studied the relation between Reichenbach's stipulation of equating universal forces to zero and the customary determination of congruence based on the use of rigid bodies and thereby divested Reichenbach's statements on universal forces of some of their misleading potentialities. Finally, he showed that the thesis that a given metric geometry determines *uniquely* a congruence class is erroneous. Analyzing Poincaré's position, he came to the conclusion that this French scientist's conception of physical geometry was not that of an extreme conventionalist but rather, contrary to general opinion, that of a qualified empiricist.

Grünbaum's thesis concerning the concept of congruence has been challenged by Putnam[192] who objected that the rigid rod is

empiricism," in *Minnesota studies in the philosophy of science* (University of Minnesota Press, Minneapolis, 1962), III, 405–526; *Philosophical problems of space and time* (A. Knopf, New York, 1963); "Carnap's views on the foundations of geometry," in *The philosophy of Rudolf Carnap,* ed. P. A. Schilpp (Library of Living Philosophers, LaSalle, Illinois, 1963), pp. 599–684.

[191] *Albert Einstein: Philosopher-scientist,* ed. P. A. Schilpp (Library of Living Philosophers, Evanston, 1949), pp. 313–332 (Robertson), pp. 287–311 (Reichenbach), pp. 665–688 (Einstein).

[192] H. Putnam, "An examination of Grünbaum's philosophy of geometry," in *Philosophy of science: The Delaware seminar* (John Wiley and Sons, New York, 1963), II, 205–255.

the ordinary standard of congruence which *defines* congruence in such a way as to establish an empirical determination of the metric, especially if perturbational or differential forces within the framework of initially not exactly known physical laws have to be taken into consideration. Putnam claimed that it is the whole system both of physical and geometrical laws, together with the correspondence laws, that implicitly specifies the metric and that it is wrong to regard the metric field "as merely a descriptive convenience which enables us to systemize the relations holding between solid bodies and clocks." In a review Smart[193] voiced similar reservations and disputed Grünbaum's contention that physical space has an intrinsic metrical amorphousness. In the defense of their theses concerning the unconventionality of spatio-temporal separations and of the ascription of an explanatory and not merely descriptive function to geometry, these authors drew the attention to geometrodynamics, which tries to present not only gravitation but also electromagnetism, charge, and mass as properties of a curved empty space alone. It remains to be seen, however, whether the new science of geometrodynamics, with its contention that metrical and topological properties of space such as multi-connectedness represent physical properties such as charge, really delivers a fatal blow to the geo-chronometric conventionalism as proposed by Reichenbach and modified by Grünbaum. The question seems to be of particular urgency since recent developments, such as the work by Sexl,[194] attempt to show that this geo-chronometrical conventionalism is merely part of a more general trend which — under the name of "universal conventionalism" — claims that any two measurements of any physical quantity at different points in space-time require coordinative definitions.

We conclude this chapter on the concept of space in modern

[193] J. J. Smart, review in *The journal of philosophy 61*, 395–401; cf. also J. J. Smart, *Between science and philosophy* (Random House, New York, 1968), pp. 246–249.

[194] R. U. Sexl, "Universal conventionalism and space-time," *Publication CTS-Phil-S-67-1 of the center for theoretical studies* (University of Miami, Coral Gables, Florida, 1967).

science with some remarks on recent investigations into the micro-structure of space. As mentioned before, in connection with Hei-senberg's introduction of a fundamental length, the interest in studying the microstructure of physical space was prompted by the desire to overcome certain fundamental difficulties in the quantum field theory of elementary particles, such as the diver-gence of self-energies due to the contributions from high-momenta virtual quanta of the proper field. As Margenau once remarked, "the degree of indulgence in speculations about discrete 'hodons' and 'chronons' on the part of physicists has been an index of the failures of their part." In fact, the early suggestions, made by Ambarzumian and Iwanenko[195] and by March,[196] even before Heisenberg, to treat space not as a continuum but rather as a dis-crete manifold, were explicitly aimed at eliminating those trouble-some infinities at any price. The resulting theories, however, proved unacceptable as they were not Lorentz invariant. One of the earliest theories of a discrete or "quantized" space-time which admitted Lorentz transformations and provided a natural unit of length has been advanced by Snyder[197] and applied by Hellund and Tanaka[198] as the basis for an operator calculus of space-time coordinates and momenta in terms of which the spectrum of the plane wave solutions of the Dirac equation could be determined.

Further progress was achieved by Schild[199] who worked with a model of discrete space-time of events (x, y, z, t) forming a hy-percubic lattice which admits a surprisingly large group of Lorentz transformations. One of its major deficiencies — namely, its impli-cation of a lower bound on the permissible velocity parameter whose value of $v/c = \frac{1}{2}\sqrt{3}$ turned out to be too large to apply

[195] V. Ambarzumian and D. Iwanenko, "Zur Frage nach Vermeidung der unendlichen Selbstrückwirkung," *Zeitschrift für Physik 64,* 563–567 (1930).
[196] A. March, "Die Geometrie kleinster Räume," *Zeitschrift für Physik 104,* 93–99 (1936).
[197] H. S. Snyder, "Quantized space-time," *Physical review 71,* 38–41 (1947).
[198] E. J. Hellund and K. Tanaka, "Quantized space-time," *Physical review 94,* 192–195 (1954).
[199] A. Schild, "Discrete space-time and integral Lorentz transformations," *Canadian journal of mathematics 1,* 29–47 (1949).

the model to ordinary physics — was subsequently removed by Hill[200] by restricting the transformations to those with rational coefficients. It is not surprising that the concept of dimensionality as defined above loses its applicability for discrete spatial manifolds. A detailed analysis of this fact has been made in 1958 by Abramenko.[201]

During the past few years some attempts have even been made, as by Tati,[202] to formulate the fundamental laws of interactions between elementary particles without using the concept of space-time, which thus becomes a statistical notion, like "temperature" in statistical mechanics, and appears as a mean value of more fundamental quantities. On the other hand, it has also been asked, and quite legitimately so, what are the limitations imposed, at least conceptually, on the possibility of measuring intervals between space-time events. In other words, to what resolution can small intervals in the space-time continuum be regarded as physically significant? A partial solution to this problem has been given by Salecker and Wigner.[203] Discarding the use of measuring rods as being essentially macroscopic objects, these authors investigated how microphysical clocks of given masses may be employed to measure distances between events which consist of collisions between particles and photons. Their results seem to indicate that quite a number of generally accepted small-scale considerations are short of physical significance. As Salecker and Wigner rightly remarked, the concepts of rigid reference frames or of (practically) rigid rods as conventionally defined and taken as the basis for the construction of the space-time metric and for the physical

[200] E. L. Hill, "Relativistic theory of discrete momentum space and discrete space-time," *Physical review 100*, 1780–1783 (1955).

[201] B. Abramenko, "On dimensionality and continuity of physical space and time," *The British journal for the philosophy of science 9*, 89–109 (1958).

[202] T. Tati, "An attempt in the theory of elementary particles," *Nuovo cimento 4*, 75–87 (1956); "A theory of elementary particles," *Progress of theoretical physics 18*, 235–246 (1957); "Concepts of space-time in physical theories," *Supplement of the progress of theoretical physics No. 29*, 1–96 (1964).

[203] H. Salecker and E. P. Wigner, "Quantum limitations of the measurement of space-time distances," *Physical review 109*, 571–577 (1958).

interpretation of the Lorentz covariance cannot be meaningfully applied in the quantal world of elementary particles. The Lorentz covariance, however, in spite of the uninterpretability of the Lorentz transformation formulae, constitutes one of the major foundations upon which the modern field theory of elementary particles is established. Thus, present-day physics is faced with a serious conceptual difficulty. It was Mehlberg's great merit not only to have pointed out the graveness of the situation, but also to have taken the first tentative step toward a solution. As Mehlberg[204] showed in detail, all existing axiomatizations of relativistic space-time, such as those proposed by Caratheodory (1924) or by Reichenbach (1924), answer the needs of macrophysics alone and do not admit a satisfactory interpretation for microphysics; moreover, they are open to a serious mathematical objection from the group-theoretical point of view.[205] Mehlberg, therefore, proposed a re-axiomatization of relativistic space-time which is not open to Weyl's objection and is formulated in terms meaningfully applicable both to macro- and to microphysics. His system consists of seven axioms which contain as their only primitive notion the concept of "collision-connectibility of two events," a relation which holds between the two events E' and E'' whenever a particle a, distinct from the particles a' and a'' in which the events occur, collide with a' and a'' at different times. It is easily understood that this concept is applicable to macro- and microphysics alike. Therefore, the proposed axiomatization offers a conceptually sound basis for a *physically* significant formulation of Lorentz covariance. A purely *mathematical* axiomatization of relativistic space-time has also been proposed recently: By axiomatizing in a direct coordi-

[204] H. Mehlberg, "Space, time, relativity," in *Logic, methodology and philosophy of science,* Proceedings of the 1964 International Congress in Jerusalem (North-Holland Publishing Company, Amsterdam, 1965), pp. 363–380; "Relativity and the atom," in *Mind, matter, and method,* Essays in philosophy and science in honor of Herbert Feigl (University of Minnesota Press, Minneapolis, 1966), pp. 449–491.
[205] H. Weyl, review in *Deutsche Literaturzeitung 1924,* 30, 2122–2128.

nate-free method relativistic space-time, Noll[206] tried to do for four-dimensional Minkowski chronometry what Hilbert had done for three-dimensional Euclidean geometry. Noll expressed the hope that thereby the study of space-time might become a branch of mathematics which, like Euclidean geometry, is apt to engage some interest from the purely esthetic point of view alone.

As the contents of the present chapter clearly shows, our knowledge of large-scale as well as of small-scale properties of physical space is intimately related to the progress in cosmology and microphysics, respectively. And as long as these branches of scientific research fail to offer satisfactory solutions to their fundamental questions the problem of space will have to be classed as unfinished business.

[206] W. Noll, "Euclidean geometry and Minkowskian chronometry," *American mathematical monthly 71*, 129–144 (1964).